T0156067

Pioneering Research

Pioneering Research

A Risk Worth Taking

Donald W. Braben

WILEY-INTERSCIENCE

A JOHN WILEY & SONS, INC., PUBLICATION

Published by John Wiley & Sons, Inc., Hoboken, New Jersey.
Published simultaneously in Canada.

For general information on our other products and services please contact our Customer Care Department within the U.S. at 877-762-2974, outside the U.S. at 317-572-3993 or fax 317-572-4002.

Wiley also publishes its books in a variety of electronic formats. Some content that appears in print, however, may not be available in electronic format.

Library of Congress Cataloging-in-Publication Data is available.

ISBN 0-471-48852-6

10 9 8 7 6 5 4 3 2 1

To the memory of my parents,
Agnes and Walter Braben

It will perhaps be well to distinguish three species and degrees of ambition. First that of men who are anxious to enlarge their own power in their country, which is a vulgar and degenerate kind; next that of men who strive to enlarge the power and empire of their country over mankind, which is more dignified but not less covetous. But if one were to renew and enlarge the power and empire of mankind in general over the universe such ambition (if it may be so termed) is both more sound and more noble than the other two.

Now the empire of man is founded on the arts and sciences alone for Nature is only to be commanded by obeying her.

Francis Bacon, 1620

Contents

Chapter 4
Taming Research: The Problems of Success

Chapter 5
The Bureaucratic Jungle

Chapter 6
Prospects for Economic Growth

Chapter 7
Re-Creating the Golden Age

Chapter 8
Venture Research

Foreword

The purpose of research is to add to humankind's knowledge and understanding—of itself and of its total physical, social, and cultural environment.

In particular, research in the physical and biological sciences is frequently coupled to new products, processes, and procedures. Here the role of research is to explore and explain that which is novel, thus extending and underpinning the platform of technology upon which developments can be based with minimum risk.

This is the central paradox of good basic research: Its pursuit is risky, insofar as it is not possible to forecast its outcome, this being the nature of research. But its purpose is to reduce the risk in whatever follows.

Don Braben recognizes very clearly the need for risk taking in both the pursuit and the support of basic research. Creative researchers must be willing to break out of the intellectual shackles of orthodoxy and should be enquirers, questioners, even out-and-out dissenters.

However, while the willingness to be identified as a dissenter from orthodoxy may be a virtue in a good researcher, it is by no means a sufficient indicator of the total capability of that researcher to be creative and to be capable of exercising sound judgment of the significance of the problem which is to be addressed.

This is where "peer review" is inevitable. But in what form? The danger in the bureaucratic evaluation of research project proposals is that it can be dominated by orthodoxy, coupled with a desire to avoid failure. Understandable when funds are limited, but hardly a formula for creative success.

It has long been my view, reinforced by an opportunity to read Don Braben's work in draft, that the way to nurture worthwhile creative research involves two steps:

1. Identify creative committed researchers (Braben's dissenters).

2. Judge whether their ideas, if pursued successfully, are likely to add significantly to humankind's knowledge and understanding base. There can be no justification for wasting valuable resources in pursuit of novelty for its own sake. That is why intelligent peer review is necessary. Braben's Venture Research initiative is one way of achieving this.

Sir DEREK ROBERTS

University College London
June 2002

Preface

I am grateful to many friends and colleagues whose support and encouragement have been invaluable. Some are mentioned here but many others are not, and I hope they will understand. I am especially grateful to my fellow board members at Venture Research International, Sir John Fairclough, Nigel Keen, and Iain Steel, who have not only been generous with their time and advice over the years even though there has been little prospect of remuneration but have also maintained an infectiously good humor and a positive outlook. Sadly, John died a few months before this book was finished. I am grateful to the many friends at University College London's Earth Sciences Department who have achieved the remarkable feat of making a physicist feel at home and to the members of the Venture Research community for their encouragement and stimulation. Peter Cotgreave, David Price, Ken Seddon, Luca Turin, and Claudio Vita-Finzi read early drafts of the book and gave helpful guidance. My wife, Shirley, and David, Peter, and Jenny provided invaluable help in the book's production in addition to their usual love and kindness.

DONALD W. BRABEN

Theydon Mount, Epping
January 2004

Pioneering Research

Introduction

Asking questions not on the agenda
Exploring ideas wherever they lead
Pursuing goals because they're important
Creating options not yet perceived

As we approach our 5 millionth birthday, it might be amusing to ask how we members of the human race are getting along. How are we managing? What are our hopes and fears? What might our future prospects be? We might also look into the fascinating question of how we got where we are today. On the planetary scale of time, we were very much late arrivals. For 99.9% of Earth's long history there were no humans, although the planet teemed with life. We did not exist. Then suddenly in the blink of a planetary eye we did. We very soon made up for lost time, however, and quickly came to dominate the place. Never before, as far as we know, has a single species made such a dramatic impact. Now, the sum of our normal day-to-day affairs has become so substantial that it is even possible we are beginning to damage the planetary home that prospered so long without us.

Divine explanations apart, what made us so different from the rest of the animals? How did we come so quickly to our overwhelming ascendancy? The picture I paint here is based on the idea that our supremacy comes from our innate dissenting trait, that powerful tendency some of us have of refusing to accept things the way they are. Ultimately, its most important expression is found in science and technology. Despite the terrible spin-offs to war and mayhem, science and technology give us the

Pioneering Research: A Risk Worth Taking, By Donald W. Braben
ISBN 0-471-48852-6   © 2004 John Wiley & Sons, Inc.

basic fabrics from which civilization is woven. We would be lost without them. If that sounds far fetched, imagine what would happen if our achievements in these vast arenas were suddenly switched off. It would be as if there had been the most cataclysmic natural disaster. We would be plunged into the tooth-and-claw existence we had to endure when we first arrived in prehistoric times. That meant fighting for our food and shelter with any other creatures that might also want them and with the elements from which we had little natural protection. We managed then because we were so few, but we have been so successful that there are over 6 billion of us today. Even our most basic needs greatly exceed what the rudimentary, science-and-technology-free resources of the planet could provide.

It is astonishing, therefore, that although the vast majority of people are willing to enjoy science and technology's copious bounties, the means by which they are created do not seem to attract much interest. If we do not like what is presently on offer, apparently, something better will soon come along: We live in an age of miracles. Yet, science and technology are almost as important to our lives as the water we drink. Life, at least for most people in the industrialized world, is pleasurable *because* of what has been done in these arenas. One must not ignore other necessities of civilization such as music, art, and literature, to mention but a few, but without the stable foundations provided by science and technology they would hardly exist. Even the earliest cave artists had to have tools or paint.

The clock cannot turn itself back, of course, and in the absence of catastrophe, the scientific harvests already gathered will probably not be lost. Indeed, they seem to have produced an embarrassment of riches, and on the face of it, technological prospects have never been rosier. Risks from war and mayhem will flare up from time to time, but most of us can do very little about that. In any case, such conflicts rarely threaten humanity's very existence. On the other hand, closer inspection will reveal a different story, even if, as we all hope, humanity's progression remains peaceful. Looking to the future, world population is still increasing. Within the lifetime of young people today, humanity will probably increase its numbers by another 2 or 3 thousand million and possibly more than that. The new science and technology necessary to maintain even present standards of living, variable as they may be, for a global population that could be 50% greater than it is now, have not yet been discovered. Can we be confident that the new knowledge essential to our future survival will be found? There is nothing automatic or miraculous in that wonderful process. Discovery is a very delicate plant. It can easily be crushed.

Even if we accept that science and technology are essential to civilized life, until recently most of us could take their precious gifts for granted. The mysterious processes of discovery could be left to look after themselves. Progress seemed to be a natural consequence of civilized life. The restless spirits among us who have so enriched our lives—the Einsteins, the starry-eyed idealists, the mad professors, the pig headed, the eccentric—did their own thing. Nobody told them what to do, but nevertheless, "they" always seemed to be coming up with some novelty or other. However, we are now seeing the emergence of trends that should be of major concern to everyone. Perhaps the most important are social in origin. In every walk of life, everybody has probably felt the consequences of authorities' current obsessions with control and so-called safety and efficiency. Their ubiquitous preoccupations are putting growing pressures on humanity to conform to specific patterns of behavior, and they affect us all. Individual freedoms are gradually being eroded. Regulations that affect every facet of life are becoming globalized. People and organizations are being obliged to be ever more cautious. Risk is increasingly seen as a bad thing to be avoided at all costs. These trends are slowly homogenizing the world. Their subtle pressures are at present no more than irritating in everyday life. In the scientific world, however, their insidious effects discourage adventurous or challenging research and curtail creativity—indeed, they corrode the very intellectual foundations on which all major discoveries have been based.

I believe, therefore, that humanity has reached a crucial point in its development, which is why I wrote this book. In the past few decades, we have seen more change than at any other time in our history. There is nothing necessarily wrong with that. My concern, however, is for its *intellectual* implications, most of which seem to have gone unnoticed. Indeed, the focus in this book is almost exclusively on the intellectual dimension— the dominance of which domain has accounted for our incredible progression and must continue to do so in the future if we are to have a future. I discuss the pathways that have led us to the present starting from our earliest beginnings. Dissent's crucial role in our progression is outlined in Chapters 1 to 3. Our civilization reached its first peak in the Greco-Roman era, but neither the Greeks nor the Romans valued experimentation. Aristotle was the most influential of the Greeks, and his extraordinary legacy lasted for some two millennia. He thought that the world was perfect because it was a gift from the gods. Being perfect, the world could not be improved, and so experimentation would be futile. Economic growth became, in the modern jargon, a zero-sum game. Once the maximum productive capacity of the land and the peoples who worked it had been achieved, there could be no further growth. None of this seemed to be understood at the time, but had the Romans been prudent they could have survived. They were not, of course, and the increasing demands from the military upon a stagnant economy, coupled with barely credible levels of greed and corruption among the ruling elite, finally destroyed them. The peoples of the west then began a slow descent into the stagnation of the Dark Ages.

The indomitable spirit of dissent, inspired by the best achievements of the Greco-Romans that fortuitously had been preserved in the more prescient Eastern countries, led to the gradual emergence of our modern (Western) civilization around about the twelfth and thirteenth centuries. Progress was fitful for three or four centuries as philosophers (later to be called scientists) struggled to escape from Aristotle's paralyzing grip. Technology had not yet been equipped with the radar of scientific insight, but early in the seventeenth century Francis Bacon was one of the first to become aware of the powerful practicable potential of science, the theoretical basis of which was transformed in that century of genius. Around the middle of the eighteenth century, the European Renaissance and the turbulent social and political events of the period created a climate of change that led to one of the most dramatic events in human history. That climate was most favorable in Britain. It stimulated a small number of talented, determined, and aggressive individuals to launch an unprecedented series of ambitious and wide-ranging technological initiatives that later became known as the Industrial Revolution. Other European countries followed closely behind Britain, and indeed Europe's technology led the world until the balance slowly tilted in favor of the United States, where it has generally remained since about the early decades of the twentieth century.

This transition was led first by untutored technology and then increasingly throughout the latter half of the nineteenth and twentieth centuries by the technologies derived from advancing knowledge.* Nowadays, indeed, although the days of the classical inventor are not yet over, the most reliable route to new technology is through the sciences. From about the 1950s, it has become increasingly clear that technical change drives economic growth, a realization that over the following few decades led to a marked expansion of the number of practicing scientists. The

*The initial development of steam motive power, for example, owed nothing to the sciences. George Stephenson could not read before he was 19 but shortly afterward was designing the steam engines that would eventually revolutionize transport the world over. It was not for another 20 years or so that this and other pioneering work led Sadi Carnot, a French military engineer, to begin to understand the rules by which heat sources can be manipulated and to initiate the development of the science of thermodynamics.

scene is therefore set for the present day. In Chapters 4 and 5, I give an outline of the almost universal arrangements that have emerged over the past 20 years or so for coping with the expansion and the simple fact that scientists' demands* for resources far exceed the supply.

These necessarily bureaucratic arrangements have created a crisis in research funding. Nowadays, it is virtually impossible for academic scientists of any status to get funded unless they can convince their peers that the work they hope to do has a suitably high priority and that they will make the most efficient use of the resources requested. The arrangements work well enough in the mainstream, but they automatically discriminate against dissent. In industry, the other main source of research funds, the main priority as ever was to look after their bread-and-butter interests, but many companies also used to give carte blanche to some of their most creative scientists to be inspired, to pursue their own ideas, and to explore as they thought fit. This apparent altruism often paid extraordinary dividends. Over the past decade or so, however, companies have moved toward optimizing efficiency and short-term competitive advantage. A common corporate policy is now "to return to core business." Consequently, companies have generally pulled out of exploratory research and have also reduced the scope of mission-oriented research in line with their more focused requirements.

In Chapters 6 and 7 I assess the possible impact of these relatively recent changes on the prospects for economic growth. Global per-capita growth has declined in recent decades. The advanced capitalistic countries of Europe and the Far East have drawn much closer to the performance of the technological leader—the United States—but the anticipated acceleration in technical progress on the global scale as more countries operate near the frontier has not taken place. The slowdown has been debated intensely, but no consensus emerges. My suggestion is that we are beginning to see the cumulative effects of the systematic stifling of dissent in science and technology. This interpretation seems to be confirmed by the fact that modern technology is based on seminal discoveries made decades ago and the flow of new sciences has been severely curtailed. Unfortunately, the popular conception today is that we have a profusion of technologies, particularly in electronics and communications, thereby creating an *illusion* of well-being.

The problems that seem to lie in wait for humanity over the next few decades are indeed formidable. Recent horrors in the Balkans, those cur-

---

*In an attempt to avoid repetition, I usually take the term "scientist" to include technologist or engineer.

rent in the Middle East and Africa, and other pockets of avoidable misery and suffering have arisen largely from extreme nationalism, racism, tribalism, or religious intolerance, but their causes also have an economic dimension. Global per-capita economic growth is declining. Per-capita food productivity too seems to be falling, after more than keeping pace with global population increases for most of the twentieth century. Nevertheless, the advanced nations seem prosperous and generally content. In these circumstances, it may seem naive or other worldly to suggest that anything that goes on in our laboratories can be more than marginally relevant.

Not everyone agrees that economic growth is necessary for our survival. Those such as the Greens say that we should learn to manage with what we have now because economic growth is equated with damage to the environment or the exploitation of people. Sadly, they are right all too often, but while greed cannot be ignored, a major source of the damage is ignorance. The more we understand and the more widely that understanding is distributed, the more likely it will be that cost-conscious organizations or desperate people will respect the environment. In any case, without growth there is stagnation, as we had in the Dark Ages. Unless human nature changes, there will always be greedy people, and they will pervert whatever system they live in. We must have economic growth to help reduce their malign influence. We have a tiger by the tail.

But the rapid growth of technology has also led some *scientists* to suggest the possibility of new types of controls. They believe that society should consider imposing limits on the new sciences and technologies that should be explored because of the risks they might pose to humanity's existence. In his book *The Final Century*, the very distinguished scientist and British Astronomer Royal Sir Martin Rees (2003) estimates that humanity has only a 50 : 50 chance of surviving the next 100 years. He cites such risks as "bio" or "cyber" terror or error that "could be graver and more intractable than the threat of nuclear devastation that we faced for decades" (cover and p. 8). He therefore concludes that we, that is, society, should place limits on the work that scientists should be allowed to do: that scientific progress should be "slowed down." I do not agree. We have, on the one hand, the *risk* of a substantial "downside" to scientific progress and, on the other, the *certainty* that if humanity does not give as much rein to its creative people as it has since, say, the Renaissance, our civilization will soon descend into a new Dark Age.

There would seem to be no reason, however, why economic growth should not only be restored to *but also sustained* at its levels of only a few decades ago or why the production of food and other essentials should not be raised to meet even worst-case predictions. This optimism is justified because economic growth receives its most important stimulation from science and technology. Until we reach the limits of scientific advance, this optimism will always be justified. Our first and modest problem is that we need to find a way of breaking bureaucracy's stranglehold on the talents of the tiny number of pioneers who are capable of making scientific breakthroughs, thereby promoting the surges of growth that usually follow.

In the final chapter—Chapter 8—I describe an initiative that set out to restore the tolerance of scientific dissent to its former levels. The Venture Research project was industrially funded and a model of corporate excellence and vision, concentrating exclusively on the most profound and basic research that even academic funding agencies would not support. It ran successfully for 10 glorious years before being sadly and abruptly terminated by an almost incredible act of corporate short-sightedness a few years before the *commercial* profitability of the ideas it spawned became a demonstrable fact.

Initiatives to restore freedom can open our eyes to new horizons and point the way to new opportunities. However, although uninhibited explorers might discover new territories, other types of pioneers must follow them if the possible gains are to be realized. Industrial leaders and investors must also be prepared, therefore, to re-create the environments within which entrepreneurs are free to back their personal judgments on risk, as they were not too long ago, and to reap the benefits or the whirlwinds as appropriate. Entrepreneurs still have this freedom so long as they do not stray too far from the beaten tracks, but we need to restore that which has been taken from them in the name of efficiency and short-term expediency.

As ever, both action and inaction have their consequences. My premise is that if we do nothing to curb the current obsessions with optimization and efficiency our future will be bleak. These twin obsessions will inevitably impose uniformity and conformity. Fashions flash around the world—not only in clothes but also in everything we do—and people in authority do not seem to have the time to assess their consequences. The mania with large-company mergers is creating commercial organizations whose value is greater than many national incomes but which nevertheless are accountable only to shareholders. The numbers of independent manufacturers of aircraft, cars, drugs, power stations, and rolling stock, for example, and suppliers of such services as telecommunications, electronic commerce, finance, and air travel are now very small and getting smaller. Competition and choice are the first casualties, but global growth too will fall progressively as capabilities converge. As I shall explain, changes to growth on the global scale are not easy to see by looking at annual figures alone. The global response time is very much longer, and for an accurate picture, population changes too must be included. One can easily be fooled by short-term spurts and regional recoveries, but for the past 30 years the inexorable long-term global trend has been downward.

To summarize my story therefore: The greatest long-term risks facing humanity will not come from such apocalyptic threats as terrible weapons of mass destruction, prolonged global war, devastating disease or famine, or extinction by a huge wayward meteor. Rather they will come from the debilitating attrition caused by the rising tides of bureaucracy and control. These trends are steadily strangling human ingenuity and undermining our very ability to cope.

The good news is that we can do something about these problems, and in principle, they should not be too difficult to solve. However, deeply entrenched attitudes can be a formidable barrier. Our major problem is that at the margin where greatness lies bureaucracy is strangling scientific creativity. Everything we value came out of the blue, out of the greatness of pioneering scientists and engineers. Such scientists as the Einsteins, driven solely by their restless spirits of enquiry, laid the foundations for a vast series of such unforeseen developments as the transistor and integrated circuitry, nuclear power, lasers and optoelectronics, and computers that today dominate everyday life. It is difficult to imagine life without them. Their work was not encouraged or stimulated. Revolutionary discoveries once emerged in the natural course of things. Human ingenuity does not need stimulating; we merely have to ensure that it is not being routinely suppressed. That vital condition was broadly satisfied from about the Renaissance until the 1970s, when the current obsessions with optimization and efficiency began. Our most urgent need, therefore, is to make the modest changes that will allow it to be satisfied once again.

1

Dissent and Research: The Supreme Stimulants

About 5 million years ago, more or less, it would seem that the great apes living in Central Africa—the hominoids—split into three distinct species. This was not an unusual event. Speciation, the processes by which living organisms suddenly branch and divide into new and ever more complex or specialized forms, had been flourishing at a frenetic pace ever since the so-called Cambrian Explosion some 500 million years before the time modern animals began to evolve. The causes of speciation are unknown, but climate, temperature, and atmospheric composition are clearly important. As life continued to expand into every available ecological cranny, that moment came when the hominoids split into the species we now call gorillas, chimpanzees, and the hominids—the earliest forms of humans. On the face of it, the hominids were just another addition to the teeming profusion of life, a late arrival to an already crowded arena. The apocryphal, all-seeing extraterrestrial may not have noticed anything unusual, as hominids at first must have looked and behaved much like many other animals. But they proved to be very different, and an apparently innocu-

Pioneering Research: A Risk Worth Taking, By Donald W. Braben
ISBN 0-471-48852-6   © 2004 John Wiley & Sons, Inc.

> ## Text Box 1: Nature
> I use the word Nature throughout this book as if it were the name of some sort of being. I am not trying to make a religious or mystical point. I use it as shorthand for the universe and every aspect of everything in it. We have made great progress, but our understanding of that system is still in its infancy. However, no understanding is denied to us. My affectionate anthropomorphism is used out of respect for a system that commands the attention of every scientist.

ous event turned out to be the birth of the species that would dominate this planet. What was it that made us so special? The competition was enormous. What gave us the vital edge? As far as I know, there are no generally accepted answers to these questions. My suggestion here is that our overwhelming advantage stems from a source that hitherto has not been generally recognized—our innate capacity for dissent. However, another condition has also to be satisfied. While a capacity for this strange intellectual ability is imbued in every one of us, our dissenting trait should normally be dormant. That is, Nature would have to arrange things so that our species has just the right quota of full-blooded dissenters—no more and no less—to liberate us from a brutish existence and to progress in a hostile world. See Text Box 1.

Some of my friends have suggested that dissent is too negative a quality with which to characterize the human race. Its negative components cannot be ignored. When our dissenting spark is transiently activated, it can lead to such disruptive behavior as violence in the classroom, the playing field, or the pub. But those responsible are rarely working to an agenda. They are merely lashing out at some aspect of their parochial surroundings. Such dissatisfaction usually has no firm foundation, and like a sandcastle on the seashore, its memory is soon washed away. The class of dissenters also includes cranks and fanatics—people with a life-long obsession with an issue and whose success is usually measured in terms of the mayhem or annoyance they create. Their ideas cannot always be dismissed as irrelevant. Many a pioneer has had to endure the slings and arrows of being outrageously cast among the lunatic fringes. And, of course, dissenters can turn out to be despots and dictators.

The form of dissent honored here, however, is the noblest of all. It is neither transient nor petulant but is usually based on an individual's overwhelming conviction that some aspect of life has become unbearable. It is the coinage of visionaries and, ultimately, the creator of hope. If dissent and the changes it brings are efficiently suppressed, all hope dies, as it did during the Dark Ages, for example, and more recently for many millions in the former Soviet Union and elsewhere. Nevertheless, even in more tolerant societies, dissent, noble or otherwise, is usually resisted or ridiculed before its spin-off might be accepted and allowed to take its place in our social or technological infrastructures. One might say, therefore, that dissent is neither positive nor negative. Warts and all, it is a neutral fact of life. It is the trait that propels us and has made us what we are. Indeed, one might say that the cumulative consequences of humanity's dissent over the last few thousand millennia define civilization.

One might expect, therefore, that with such wonderful achievements to its credit, our defining trait would be celebrated and protected. That is not the case, nor is it necessary that it should be. Indeed, despite the fact that we owe it everything, our crucial characteristic has always had a relentless and implacable enemy—bureaucracy. Bureaucracy's roots, under whatever name, can be traced back to pre-Roman times, but pioneers have learned to live with it. However, in recent years, its power has grown enormously. Aided by modern electronics and communications and festooned with regulations that are correspondingly easy to enforce, it is almost invincible. Nobody seems to like it, but its power grows anyway. Ironically, the most important problems arise in the scientific world when the bureaucrats holding the purse strings strive to be fair to all-comers. This counterintuitive conclusion arises because even-handedness is often a policy for ducking the tough decisions. Fair shares for everyone or moving forward by general agreement are easily defendable policies. In some fields of human endeavor, they may even offer the best options, but science and technology are not among them. The discoveries that have transformed our lives often came from scientists who defied consensus. The goals they set themselves were not always reasonable. They often challenged what was thought to be possible. Until recently, there was just enough slack in the system to allow such pioneers to flourish. That is *exactly* how it should be. Nowadays, the bureaucrats have closed these loopholes on the spurious grounds of efficiency, and pioneering projects will probably be set aside as a risk too far, especially when funds are short.

It has only recently been formally recognized that science and technology drive economic growth (see Chapter 6). Nevertheless, the metaphorical engines that drive the processes of growth are still rather mysterious. Economists would rather concentrate on fiscal issues because they are thought to be well understood and the impact of fiscal changes can be almost immediate. Unfortunately, tough problems have a habit of biting back if they are ignored or misunderstood, and everyone can be affected.

On the wider scene, the threats from such diseases as AIDS or cancer or the problems of population growth or global warming are obvious examples. In these cases, the problems seem intractable, but scientifically at least we are doing what we can. However, it has now become fashionable to tinker with the engines of growth even though they have been working well for more than a century. In the mundane world, most drivers are smart enough to abide by the if-it-works-don't-fix-it rule when it comes to real engines. Economists and those who take their advice should take note.

Unfortunately, such a let-it-be policy is not consistent with current fashions. Everyone knows that scientific advances come from inspired scientists and engineers—from rare and wondrous geese that lay golden eggs. In the past, they were allowed to roam freely and create as the mood took them. Their success reached its peak during the 1950s, 1960s, and early 1970s—a period economists have called the Golden Age because global economic growth was so high. Nowadays, however, it seems to have been forgotten that unless highly creative people are given total freedom, they will probably not realize their full potential. Indeed, the central theme of this book is that humanity will continue to prosper only if we allow pioneers to be no less free than they have been throughout history.

I am optimistic that we will overcome, but optimism comes in many shades. At one end of the spectrum, those people who are disposed to the blind variety do not care what fates await them. They *know* that they personally, or the people in whom they put their trust, will overcome. Then somewhere in the middle there are those who declare that they are slightly, somewhat, or rather optimistic, and one waits for the conditional "but" that will give voice to their pessimism. At the other end, a small number of what I will call professional optimists will be found. These Panglossian characters have their eyes wide open and plan their positive strategies no matter how appalling the dangers that might lie ahead. Most people seem to be confident that humanity's boundless ingenuity will somehow find a way forward through the apparently intractable problems that face us. Professional optimists, however, are not idle philosophers— they are prepared to take the actions implied by the specific grounds for their optimism and to pursue them to the limit of their resources. This book has been written by one who strives to be such a professional.

Judging by the catalogue* of awfulness that lies ahead, optimism seems to be essential to survival. The inexorable rise in population seems

---

*A thorough review and a compendious bibliography may be found in Paul Kennedy's (1993) *Preparing for the Twenty-First Century.*

to be one of our most important problems. In this respect, humanity's turbulent progression is entering a new phase. For many centuries, our imaginary extraterrestrials looking down on us might have thought that our numbers were largely predetermined. Long-term fertility rates barely overcame the natural ravages of famine and disease, and the unnatural tolls of war. Consequently, between about 8000 B.C. and the beginning of the eighteenth century, global population hardly increased* throughout a healthy human lifespan. For men and women struggling for existence, the local ebb and flow of life-threatening forces would often have been overwhelming. This is still the case in the poorest parts of the world, but even the supposedly well-managed rich countries can now feel the pressures arising from their modestly increasing populations.

Optimism will not be enough, of course. We must recognize that the quest for ever-higher levels of efficiency also has its downside. The more efficient we make our institutions, the less flexible they become. There is no process in Nature that is 100% efficient, and the closer we try to encroach on that unachievable limit, the more obstinate Nature becomes. For human enterprise, maximum efficiency usually means pulling together, which is fine if everyone agrees on what we should be pulling toward. Hopefully, that will be the case for most of the time. We humans are very gregarious. The dangerous problems arise when we push the envelope of efficiency too hard. We humans did not get where we are by moving unquestionably together like the common herd.

But I am getting ahead of my story. In this and the next few chapters, I would like to say something about how our present precarious state came about before turning later on to what we should be doing about it. Research—the quest for new sciences and new ways of doing things—has been the instrument of humanity's outstanding success. But before amplifying on the essential role played by dissent, we should be clear about what research is. In everyday life, criminals believe that laws are made to be broken. Weakness in the law or in its enforcement should be exploited for the maximum personal advantage. If crime rises, legislators will normally respond with more enforcement or more law. If the changes prove to be effective, the law abiding can usually be persuaded to offset the additional restrictions against the better security of person or property. But laws curtail personal freedom, and few are happy with the law's increasing scope.

---

*The average trend of global population increase throughout the centuries leading to the Industrial Revolution was about 0.04% per annum (pa)—i.e., some 4% a century.

Scientific laws, however, work quite differently. If a new law or a fixable weakness in an accepted law is discovered, the results will probably be beneficial. We reap this reward because scientific laws are statements on how we believe Nature works. The more we understand, the more we will be able to persuade Nature to help us on our way. Since Nature's complexity seems boundless, the hungry scientist should never lack for motivation or employment. However, researchers should be realists. While they may take pleasure in exposing flaws in established scientific foundations and putting them right, they must recognize that it will only be a matter of time before their reinforcements are found wanting by smarter searchers after truth. These restless pursuits have no downside, however. We all gain because replacing ignorance with understanding is very rarely a bad thing.

And so the restless quest for new knowledge has gone on since fully fledged humans made their appearance on the global stage more than 50,000 years ago. There were no laws of any kind in those days, except perhaps that of the survival of the fittest. Hundreds of thousands of species have made their appearance on the planetary stage over the geological times since life first emerged. It seems that every possible ecological niche that will support life, from the poles to the ocean depths, eventually came to be occupied. However, each species tacitly agreed, anthropomorphically speaking, to make the best of whatever they could get from the flora and fauna that Nature provided. There could be no complaints, and they would prosper or perish according to their ability to adapt. But then humans suddenly arrived, and we took a very different line. At some crucial moment in the distant past, the creatures that became our ancient ancestors became seriously dissatisfied with the pact that their fellow species had always accepted. This was not a passing phase. It did not fade away when such problems as hunger or homelessness had been solved, as it does for other animals. Our defining moment seems to have created a determination not merely to succeed—all animals have that—but to enhance and extend the ranges of success. Such determination had not been seen before. It led to us becoming the dominant species.

I have focused on dissent because otherwise we do not seem to have much going for us. Many animals are better than we are in some way that could make an important contribution to eventual dominance. We are not the strongest or the heaviest and most certainly not the largest (see Figure 1). The old one-liner—Where does a 10-ton gorilla sit? Anywhere it wants.—applies only in a joke world, apparently. Cheetahs, among many others, are faster; horses have more stamina; eagles better eyes;

Figure 1
Size is clearly not important when it comes to determining the capacity for lasting world dominance. The skull of *Carcharodontosaurus saharicus*, the largest of the predator dinosaurs, was discovered in 1995 by Paul Sereno and co-workers in the Sahara in Northern Africa (*Science*, May 17, 1996, p. 986). Dated to 90 million years before present, its brain size is only about one-fifteenth that of an average human today. It is shown here with a modern human skull on the same scale. But, of course, humans and *C. saharicus* were never in competition as the dinosaurs became extinct some 65 million years ago. (Image reproduced by permission of Paul C. Sereno, University of Chicago.)

bats better hearing; dogs a better sense of smell. We are relatively poor swimmers. We have no natural weapons such as horns, fangs, or claws. Other animals can coordinate their movements at least as well as humans. They can also cooperate—hunting in packs, for example. Humans can communicate, but we are not alone in that, and it would seem that primitive human vocal chords were poorly developed so our communication skills would not have given us much of an advantage to start with. Nevertheless, we became the undisputed number 1. Why?

The all-important edge could have stemmed from intelligence. But intelligence is not spread evenly among a population. Its natural variation—sometimes popularly referred to as the bell curve—means that a tiny proportion of humans can be exceptionally brilliant while a similar proportion may have severe learning difficulties. Animals are also thought to be intelligent, and since the spread of intelligence among members of each species will have its own natural variation, it is possible that some exceptionally gifted individual animals will be more intelligent than some disadvantaged humans, however uncomfortable that prospect may be. Those especially gifted animals would be generally more successful. As even slight advantages can accumulate to yield enormous benefits over many generations, evolutionary pressures should have increased the average intelligence of that species with time. If it has, it does not seem to have changed ways of life. As far as we know, animal behavior—humans excepted—has been unchanged for millions of years.

Intelligence, however, is not easy to define or to measure. The conventional measure—the intelligence quotient (IQ)—usually selects for abstract reasoning ability, but people with very high IQs can sometimes be hopelessly impractical—too clever by half. Since, in addition, intelligent animals do not seem to have progressed, it is not easy to see how intelligence alone could determine the triumph of a species. In any case, primitive human brains seem to have been much smaller than those of modern humans. They may not have been especially intelligent therefore. Intellectual capabilities seem to be very broad, however, and they include another ingredient that seems to have been vital to our success—the capacity for dissent.

Many of our animal functions are automatic. We do not have to remember to breathe, to blink, to digest what we eat, or to activate our reflex actions. Functions like these are preprogrammed, as they are for animals in general. A natural inclination to dissent, however, seems to be confined to humans. "Oh no its not," some might say. Of course, animals fight, as we do over food, mates, territory, or anything else that might affect our immediate future. However, what I am concerned with here is not anger or stubbornness but the dissent that leads to principled, determined, and sustained action. This type of dissent seems to be related to the scope for individuality. The cartoonist Gary Larson seems to have expanded on that idea in some of his *Far Side* cartoons. They are funny because he often places animals in preposterous situations. Replace Larson's mischievous animals by people, and the humor would disappear. Animals do show individuality, but it is limited. Animals that live communally are generally content once the pecking order has been established. Their leaders come and go, but pack or herd dominance is their sole objective. Once leaders have made it, they never try to change the way their followers live or how they behave.

But humans do. Leadership is not enough. Following our somewhat miraculous event, whatever may have been its cause, we began to do new things. We created such artifacts as tools, weapons, clothing, and shelter. Some animals do some of these things, of course. Primates use sticks and stones. Birds build nests. Ants and termites go in for elaborate structures. And so on. But humans had an extraordinary capacity for individuality. There always seemed to be someone who would be very unhappy with some aspect of our lot. These dissidents did not stop there. They were not merely whingers—they had to change something, to leave their mark. They might not have known quite what to do, and there would probably have been a great deal of thrashing about. It would be presumptuous to call it research at that early stage in our progression, but there was trial and error, not only with material things but with our social structures too. Some of them stuck. Step by tiny step, therefore, foundations were laid down that would lead to a radically new way of life, foundations that could be built on inexorably, generation after generation. Our behavior gradually changed. Eventually, when modern humans finally did emerge, we were so different that it would be possible to believe that we were not really animals at all. We became people.

There is no general agreement on the causes of this miraculous transition. Perhaps it was because we had extended the processes of evolution further into the intellectual domain than any other animal species had done previously. Hitherto, animals had progressed or perished dependent on their fitness for purpose determined mainly on the relationship of their physical abilities with the environment. Each species had its specialist niche, but its prosperity was also determined by its ability to cope with change. Thus, polar bears thrive in the Arctic but would not survive a tropical climate. Dolphins would not survive on land. Our species shared many of the physical advantages of the primates from which we evolved, and these abilities must have played an essential role in the early days of our separation. But those abilities also included the scope for intellectual expansion, and we moved so far into this intellectual domain that we found we had no competitors there.* We had found virgin "territory," so to speak. We could rule the roost without challenge.

The potential of our precious discovery would seem to be almost limitless. Our new type of fitness can be used to allow us to survive anywhere—even outside our planetary home if we so choose. Problems that might be fatal to the survival of other species can be solved. We can quite literally move mountains. But there is a snag. We must continually

---

*One might offer the tongue-in-cheek but nevertheless terrifying conjecture that our future excursions into such areas as supercomputation and artificial intelligence may eventually create intellects more powerful than our own!

do our best to maintain the conditions that allow the dissenting few to have their say. They do not need encouragement, but neither should they be actively suppressed as they were in the Dark Ages, for example. Today, we have a new problem in that we have become obliged to believe that dissent is synonymous with inefficiency, however miniscule the dose. But I must not get too far ahead.

Our separation from the primates* did not happen overnight. Indeed, it was a very protracted affair. It took ages because the very concept of purposive change had to be developed and accepted. As far as we know, that concept does not exist for other species. They are driven by the need to survive, of course, but other species do not seem able to respond to say radical changes in the environment or the food supply if they exceed certain limits. Humanity progressively and purposefully extended those limits, but we had to learn how to do it. As we know today, most people usually resist change. Routine and the status quo can offer comfortable ways of life. Things usually have to get worse before they can get better. When they become unbearable, we tend to be receptive to the idea that something must be done, thereby creating fertile ground for the persuasive dissident.

It is probable, therefore, that even such important inventions as the wheel and the bow and arrow would have been adopted only when everyone agreed that they offered overwhelming advantages. Thus, it would seem that the fomentation of enduring change needs a well-defined launching pad. If our ancestors had listened to everyone who tried to change our ways, we would soon have sunk into chaos and instability. As if by magic, therefore, Nature's evolutionary forces seem to have endowed humanity with just enough of the spark of our special type of dissent to ensure that we separated from the animal kingdom and prospered. The changes it led to would have to be significant to create a new species. On the other hand, their effects must also be slow and subtle. According to current understanding, some 5 million years of their evolutionary pressure was required to forge from the hominids the beings we can now clearly recognize as primitive versions of ourselves.

I do not mean to imply, however, that evolution is smooth and orderly.[†] On the contrary, it seems to go in fits and starts. Change is often succeeded by long periods when nothing much happens. *It is as if* there were indeed evolutionary forces that work in close collaboration with

*Strictly speaking, our species is of course included within the primate order. We are still part of it.
[†]This subject is still very controversial. For an elegant and short overview, see the book reviews in *Nature* by Michael A. Goldman (2001, p. 252).

similarly magical forces managing the land, sea, and air environments. These forces seem to take care to consolidate each change and to check that it has produced a fitter product before starting the next round. And so on. The nature and distribution of our dissent too had to be regulated. Without our dissenting spark, we would not have escaped from the tooth-and-claw existence that other animals must still endure. On the other hand, if the spark had been distributed too freely, social cohesion could not have developed. Progress would then be impossible as there would never have been sufficient numbers to agree on what should be done or to consolidate a new status quo. Chiefs need Indians.

I must not get carried away. Evolution is not preplanned, and there is no such thing as destiny. The famous American physicist Murray Gell-Mann once said that Nature is a purely totalitarian state. Everything that is not forbidden is compulsory. That is, everything that can happen will happen.* The "forces of evolution" are not needed, therefore. But it does no harm to anthropomorphize on the reasons why we might have emerged. Such simplification might even help us to stimulate an interest in how we might survive in the future.

Limiting the dissenting trait to a light sprinkling would mean that the trait would normally be dormant for most of us. This parsimony led to a remarkable consequence. Turning for a moment to less abstract considerations, we can take comfort from Nature's success in creating immune systems that generally protect us from such alien bodies as bacteria, viruses, and toxins and keep our bodies healthy most of the time. It would seem, however, that our intellects have also evolved a form of protection for acquired knowledge, whatever its quality. Thus, what we *know* becomes a part of us—a possession that must be kept safe from all challenges. We have all met those infuriating stick-in-the-muds who doggedly refuse to change their ways even though the case for change can be overwhelming. Some of us might even admit to similar obstinacy ourselves. But immunity from doubt could have played a vital role in human evolution. Humanity needs its dogged determined narrow-minded dinosaurs to defend their ways of life. Pioneers *should* be made to fight for what they believe in. Our ancient ancestors would not have wasted their time arguing about such questions as who had the best techniques for lighting fires, skinning animals, or simply staying alive. These skills would be passed down the generations and defined the status quo. In contrast with the rest of the animal kingdom, however, our light dusting of dissidents

---

*The ancient Greek philosopher Democritus is credited with a similar remark: "Everything existing in the universe is the fruit of chance and necessity."

could sow the seeds of permanent change. It was a very slow process, but we gradually ratcheted ourselves away from the animals as we moved further and further into the intellectual domain. The animals remained wild, of course, but we grew increasingly confident about our collective abilities. It turned out to be a tender trap. Humanity became the first of the tamed tellurians—*the first animals on this planet to domesticate themselves.**

The recipe for creating the human race seems to have been simple. Take a suitable species, particularly one with a capacity for intellectual development. Give it a temperament prepared to accept the idea of purposive change. However, let that species be superbly confident and generally untroubled by self-doubt. Then, spice it with a pinch of creative and determined dissidents. Let it simmer slowly for about 5 million years. If the species is generally complacent, it will mean that anyone who tries to change the status quo will meet with stubborn or hostile resistance. But those rare and troubled individuals in whom the dissenting spark has been fully ignited will not easily be deterred—the instigation of change is their consuming passion. Many might fail, but others will follow. Eventually, the recipe ensures that the species passes through the doors of domestication without protest. The species will have become civilized. The human race will also be ready for anything.

Those who would bring about lasting change will usually need to understand the issues involved. The concept of research should therefore be intrinsically attractive because it provides the means by which our dissent can be made more effective. Research, however, involves a great deal more than mere idle curiosity. Many animals, particularly when young, engage in playful exploration and experiment, but any sense of wonder is fleeting, it never becomes a consuming interest, and it does not lead to permanent changes in their way of life. It is surprising, therefore, that when the time came to give a scientific name to our species, Carl Linnaeus (1707–1778), a Swedish botanist, chose the Latin *Homo sapiens.* It means "wise man," and it was not a good choice. Wisdom is too passive a trait to explain how we came to our present dominance. We are certainly not born with it. Wisdom must be acquired—usually painfully, if at all. In the young of other animal life forms—lions, gorillas, birds, or fish—it is possible to see the signs of their major characteristics soon after they are born. Children, on the other hand, have many wonderful ways, but being wise is not one of them. Passionate and raucous dissent seems

*Many types of insects—ants, bees, etc.—lived in social harmony long before we did. However, their behavior seems to be programmed and unchanging. As far as we know, they live today as they lived millions of years ago.

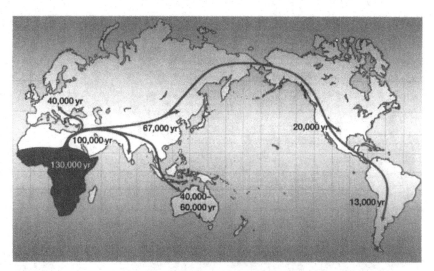

Figure 2
The origin and dispersal of modern humans as estimated from mitochondrial DNA samples by S. Blair Hedges of Pennsylvania State University. Our time of arrival as a distinct species is not well known, but Blair Hedges estimates that it was in the range 130,000–465,000 years ago. (Reprinted by permission from *Nature 408*, 653 (2000). Copyright 2000 by Macmillan Publishers Ltd.)

to come naturally to us all at first; it is only later that most us become tamed for most of the time.

Linnaeus was one of the most perceptive scientists of the eighteenth century. To understand the thinking that led to his remarkable choice, we have to go back to the time of the first disciplined scientists—the ancient Greeks. Aristotle, who lived from 384 to 322 B.C., was perhaps the most influential of them all. He believed that all living things could be classified into two distinct kingdoms—plants and animals—and each had its place in a hierarchy. In the plant kingdom, he placed the algae and the ferns at the bottom, and the more prestigious levels were given to the higher plants, including those with flowers, seeds, and fruits. In the animal kingdom, the possession of a backbone was the determining factor, and not surprisingly, since a member of our species was awarding the grades, he placed humanity at the highest level.

One cannot make progress in scientific research without making assumptions. They are analogous to the use of pitons, karabiners, or ropes in a difficult mountaineering climb, and one's progress will depend on

Figure 3
Isaac Newton's birthplace—the manor house of Woolsthorpe, near Grantham in Lincolnshire. Newton (1642–1727) was the son of a farmer. (Photograph by the author)

their strength and durability. The convention in research is that any assumptions should be clearly stated alongside one's discovery. Thus, one might claim, as Isaac Newton could have done to illustrate his law of gravity, that projectiles travel in a curve (a parabola) determined only by their initial velocity and Earth's gravity. Assuming air resistance can be neglected, the projectile's mass and shape would not be important.

Among the most perilous pitfalls in research, however, are the assumptions arising from subconscious prejudices. They might arise from an unquestioning acceptance of common sense—the generally accepted wisdom* of the time—but all too often, it is not sense at all. In Aristotle's case, he took as incontrovertible fact, as everyone did in his day, that the

---

*In his wonderful book *Tools for Thought*, C. H. Waddington (1977) introduces the concept of Conventional Wisdom of the Dominant Group. As he says, "If you would like to contract this lengthy phrase in the fashionable way, COWDUNG is memorable, appropriate, and accurate enough" (p. 16).

Figure 4
Jeff Kimble, a physicist from Caltech and a Venture Researcher (see Chapter 8),
"delivering" Newton into the world from his very bed in the manor house.
(Photograph by the author)

species inhabiting Earth were distinct and immutable. They were the same
when Aristotle was alive as they always had been. They would never
change because they were the product of divine creation.

Linnaeus was the son of a country curate, Nils Ingemarsson, but
Swedish families in the eighteenth century did not always use family
names. His father chose Linnaeus after a beautiful Linden tree in the
family garden. He used the Latin form after the fashion of the time and
because Latin was the language of the scholars his father hoped his son
would soon join. Linnaeus began his training at the University of Lund in
1727, but with an eye on earning his future living, he initially chose medicine rather than his beloved botany, which had little more status among
scholars of that time than a hobby. In the following year, he moved to the
more prestigious University of Uppsala, where he must have been surprised to find that his botanical expertise was quickly appreciated. He
won a commission from the Academy of Sciences to travel throughout
Lapland, a journey that led him to develop his ideas for classifying plants.
Such a system was urgently needed to bring some order to the profusion

of discoveries being made at the time. How could one claim to have found a new animal or plant if it cannot be described in a language that will be widely accepted and understood?

On his return, he met a fellow medical student, Petrus Artedi, who also had a passion for natural philosophy. They soon realized that their skills and temperaments were so perfectly matched that they might tackle something important. Neither lacked ambition, and while they were both still in their twenties, they made a pact that they would classify the known organisms in the entire living world, no less. Later in my story, I will relate how nowadays the ways by which research is administered and funded make it very difficult, to say the least, for such youthful exuberance to flourish—how the various research bureaucracies will insist on being persuaded that an impossible mountain in all but the eyes of dauntless youth is not only worth climbing but it can be proved that it *can* be climbed; that a more manageable peak in a more accessible range might not be a better choice; that the young pretenders are the best people available to make the attempt; that the cost cannot be reduced; that they have worked out a realistic route to their ambitious summit taking into account every possible hazard they might meet; and that they have devised a credible timetable for reaching each milestone along the way. All this red tape must be neatly tied up and approved by the bureaucrats before they would be allowed to take the first step.

Fortunately, their breathtaking audacity was matched by their energy and dedication. The only obstacles they had to overcome were those created by Nature herself, and she is never bureaucratic. They started by carving up the living world between them according to a simple formula. Linnaeus did not like fishes or reptiles, so Artedi took those: Linnaeus assumed responsibility for the rest. Their collaboration turned out to be very fruitful and survived their move to Holland, where Linnaeus took his formal medical examinations. Tragically, after they had been working together for only five years, Artedi drowned when he fell into an Amsterdam canal at night. Such was their meticulous attention to detail that they had agreed that if one of them died, the other would take over. Linnaeus therefore continued with the momentous task alone. He usually worked at a frenetic pace, but nevertheless it took him another 25 years.

For Linnaeus, species of organisms could be grouped into higher categories called *genera* (singular, *genus*) and further classified by a name for the *species* that expressed a characteristic. Aristotle had used a similar scheme, using the word genus for a group of similar organisms together with the *differentio specifica*—the specific difference of each type of organism. But there was no agreement on how genera should be grouped.

Groupings were often arbitrary, lumping, say, domestic or water animals together because they had a similar habitat. Linnaeus's innovation was the grouping of genera into higher *taxa* that were also based on shared similarities. In Linnaeus's original system, therefore, genera were grouped into orders, orders into classes, and classes into kingdoms. Thus, the kingdom Animalia contained the class Vertebrata, which contained the order Primates, which contained the genus *Homo* with the species *sapiens* expressing the characteristic wise.*

Linnaeus's scheme simplified naming considerably. He designated one Latin name to indicate the genus and another that could be used as a shorthand name for the species. The binomial system, as it came to be called, was first established for the plants and was published in its final form in 1753 in his *Species Plantarum*. It was followed five years later by the final edition of his *Systema Naturae* that covered the animal world. Hitherto, the animals and plants had been described by whatever features the observer thought might be important, but the resultant shambles made it difficult to recognize any order there might be among the many thousand types of plants and animals known even at that time. Linnaeus's succinct shorthand, on the other hand, was easy to use. It came to him, as he described, "like putting a clapper to the bell." The binomial system not only appealed to contemporary scientists but has also proved to be so powerful that it has continued in use to the present day (with modification). It also caught the public's imagination. Anyone could now name the plants and wild flowers they might come across and enjoy the satisfaction of placing them accurately in Linnaeus's great scheme.

Linnaeus was a devout Christian. Like Aristotle, he strongly believed that all living things had been created at some divine moment in the profusion that we see them today. In the somewhat pompous language of the time he saw organic Nature as an "endless cycle of life and death in which every plant and animal fulfils its destined task in the service of the whole." Although he came to modify these views slightly as a result of his work on plant hybrids (which of course as any gardener knows cannot be divine creations), the seeds of doubt did not take root. As I shall outline later, Aristotle's thinking exerted a virtually paralyzing effect on more than *50 generations* of philosophers and scientists and was still exerting its intellectual grip in Linnaeus's time. As might be expected, therefore, Linnaeus attributed such aberrations as hybrids that he had actually observed with his own eyes to "the work of Nature in a sportive mood." He

*Linnaeus originally gave humanity the name *Homo diurnis*, as opposed to *Homo nocturnus* for the orangutan.

remained a firm advocate of "the fixity of the species" until the end of his life in 1778, believing that he was a prophet called by God to promulgate "the only true dogma" and that any who disagreed were "heretics" and deserved to be persecuted.

Some 50 years later, when he was only 29, Charles Darwin (1809–1882) formulated his ideas on the origin and evolution of the species. However, Linnaeus's reputation as a brilliant observer weighed heavily on him. Many great discoveries often appear obvious with hindsight. Indeed, after reading Darwin's eventual publication, Thomas Huxley famously remarked, "How extremely stupid not to have thought of that." But Darwin had to ask himself why the great Linnaeus, who apparently missed nothing, had not thought of it. He also feared, quite rightly, that his ideas would be fiercely opposed by the scientific and religious establishments. He was also concerned that they might offend his devout close friends and family. Not surprisingly, therefore, he sat on them for 20 years. In 1858, however, he was shocked to receive an essay from an up-and-coming youngster—Alfred Wallace (1823–1913)—who had independently conceived similar ideas on evolution. Rather than rush into print to squeeze each other out, as the competitive climate would virtually enforce today, they showed a wonderful spirit of camaraderie in agreeing to publish jointly.* Ironically, they chose the *Journal of the Linnean Society*, set up in London[†] to honor the great man. Their momentous paper swept away much of the dogmatic thinking that Linnaeus and others had so passionately defended. It also created a furious controversy that lasted many years.[‡] See Text Box 2. The study of biology subsequently grew explosively. It also stimulated growth in such other fields as geology, physics, and cosmology that might help us to understand the origin of life—"the mystery of mysteries"—as it was called in Darwin's time, a description that would not be out of place today.

The convention in biology is that once the holotype (the original

*They remained friends for the rest of Darwin's life. Wallace was a pallbearer at Darwin's funeral.
†When Linnaeus's son died without heirs, his wife and daughters sold Linnaeus's library, manuscripts, and natural history collections to the English natural historian Sir James Edward Smith, who founded the Linnean Society of London to take care of them.
‡Even politicians joined in. The Earl of Beaconsfield (1804–1881, i.e., Benjamin Disraeli) remarked that it seemed to be in dispute whether men were descended from apes or angels, and that for his part he was "on the side of the angels."

Text Box 2: Evolution and the Church

It was not until 1996 that Pope John Paul II acknowledged the existence of Darwin's concept of evolution. (Sadly, Wallace's contribution is rarely mentioned today. His first full-scale biography—*Alfred Russel Wallace: A Life* by P. Raby—was not published until 2001.) The Pope indicated that the Catholic Church was ready formally to accept scientific evidence that evolution is more than just a hypothesis, saying that it is acceptable to believe that "the human body originates from living matter which predates it." He went on to say, however, that Darwin's view that "the spirit is also a product of matter" was unacceptable, as it would lead to an irresolvable conflict between science and faith. The influential Creationists—particular in some states of the United States—resist Darwin and Wallace's ideas to the present day.

specimen from which the description of a new species is established) of an organism has been deposited in a national library or museum and its name has been accepted, it should not be changed. For plants, the holotype is usually a pressing of the dried plant. For animals, they might be stuffed or preserved in part or whole. For human beings, this treatment is not now possible, although Jeremy Bentham (see Figure 5) eccentrically opted for it in his will. However, it has been jokingly remarked that the holotype of our species was eventually deposited by Linnaeus's family in 1778 at a site 6 feet below the ground in Uppsala, Sweden.

As I have mentioned, however, the name he chose for us—*Homo sapiens*—leaves much to be desired. It was in Latin, of course, the language of scholars at the time. Today, the equivalent language is English (or American), and my choice for our species would focus on humanity's capacity for dissent. Thus, it would be *Dissentient man* or *Homo dissentiens* if Latin must be used. It is not likely to be accepted. However, I suggest it because it could play an important role in our future survival. It could provide a constant reminder that our species has reached its present ascendancy only because successive generations have always spawned a sprinkling of dissidents who would challenge the generally accepted wisdom. Indeed, we will survive only if we continue to change. *Homo sapiens*, on the other hand, encourages the idea that our privileged position is unassailable. Wisdom does not necessarily have to be refreshed and renewed. Once one has been accorded the status of being wise, one can sit back and enjoy it.

Figure 5
One of the most influential people in the founding of the university at which
I am a visiting professor—University College London (UCL)—was Jeremy
Bentham (1748–1832), one of the nineteenth century's most radical philosophers.
Thanks to Bentham and others, UCL has a tradition of tolerating dissent and
was indeed the first university in England to accept Catholics and other
nonmembers of the Church of England. Bentham had his body preserved and
installed in the main entrance to the college, where seated amiably in a glass case
he could greet all comers. The cabinet contains Bentham's preserved skeleton
dressed in his own clothes and surmounted by a wax head (his mummified head
is kept in the college vault). His welcoming effigy marks a very convenient
meeting place. (Photograph by the author)

In fact, our position is delicately balanced. As the last century turned,
the euphoria helped hide the fact that we were facing an avalanche of
serious problems. Our only hope for the future is that the complacency of
the majority does not overwhelm the efforts of the dissenting few who can
keep us one step ahead of the game.

2

The Power of Dissent: From Primates to Superpower

Humanity's progression has peaked many times during the past 10,000 years. Civilizations such as the Sumerian, Egyptian, Cretan, and Chinese dominated huge empires for a time before they ran into the sands. Eventually, the Greeks and Romans brought us to new heights, but their civilization too was relatively short-lived. Astonishingly, that brilliant period of enlightenment was followed (in Europe) by centuries of stagnation and self-satisfied ignorance. For generation after generation, the lives of ordinary men and women could again be characterized as being poor, nasty, brutish, and (thankfully) short. But we recovered, of course. Historians usually present these ups and downs as chance events. Civilizations rise or fall depending on the arrival or downfall of a great ruler or a new system of government. Unpredictable natural disasters such as crop failure or disease also take their toll. My thesis is that these apparently random cycles have a deeper origin. Civilizations flourish when humanity's ad-

Pioneering Research: A Risk Worth Taking, By Donald W. Braben
ISBN 0-471-48852-6 © 2004 John Wiley & Sons, Inc.

venturous spirit is allowed to burn brightly. We falter when authority strives successfully to snuff it out.

There is no need here for a history of humanity—they are available in abundance. But traditional histories usually concentrate on social or economic intrigues, conquest, and general mayhem. I want to take a different perspective, but the subject is so vast that I can do no more than sketch an outline. Scientists, however, often take pleasure in making simple calculations on prodigious problems, typically using whatever scraps of paper that come to hand. These rough scribblings are often surprisingly accurate. Indeed, major exploits often start out that way. I would like to offer, therefore, a back-of-the-envelope history of our *intellectual* development to the present day. Its perspective must be sweepingly panoramic, but hopefully it may inspire others to review the ways we run things.

It is not even clear when the history should begin. Some experts take our separation from the Neanderthals as the starting point, beginning some 50,000 to 200,000 years ago.* But we have been evolving for much longer than that, and as far as we know, we still are.[†] Large numbers of paintings found deep in caves at Ardeche and Lascaux in France and at Altamira in Spain, among others, have been dated to about 28,000 B.C. These masterpieces would grace any modern art gallery. Little is known of the people who painted them or why they did so. But they are the first recorded attempts by our ancestors to represent images of their experiences, a vital first step on the road to writing. They would also seem to indicate similar intellectual abilities to those of our own today. Indeed, one might imagine that an infant transported from those early times would probably develop quite normally alongside a modern child. However, we know almost nothing about the rate at which intellectual potential evolves. If artistic ability is a good measure, evolution is so slow that 30,000 years does not seem to make much difference. Our almost imperceptible progression would therefore seem to favor a much earlier beginning. Perhaps *Australopithecus* (southern ape) was the stock from which

---

*The type specimen Neanderthal skeleton was discovered in the Neander Valley near Düsseldorf in Germany in 1857. Based on the analysis of mitochondrial DNA from the original skeleton, the evidence is growing that the Neanderthals were a distinct species that separated from the lineage that eventually gave rise to modern humans about half a million years ago. It seems that they might have coexisted, at least in Europe, until about 40,000 years before the present B.P. See the review by Paul Mellars (1998).

[†]The time, place, and origin of human evolution are very contentious. However, Ernesto Abbate et al. (1998) reported the finding of a well-preserved *Homo* cranium in Eritrea which they dated at 1 million years before the present.

we sprang, a primate that lived some 4 to 5 million years ago? If our imaginary magician first ignited the spark of dissent at about that time, it soon led to the emergence of *Homo erectus*, our first ancestor to walk in much the same way as we do.

Having cavalierly disposed of our first few million years, we can turn to the last few hundred thousand.* Standing on two feet stimulates face-to-face communication, and so spoken language can begin to evolve. This development seems to have triggered cranial development. It led, in turn, to further cultural progress, but it seems to have taken more than a million years for our brain to expand to its present size. Culture—the achievements, customs, and civilization of a particular time—seems to have been the key to the expansion of the human mind and eventually to brain enlargement. But progress was excruciatingly slow, perhaps because our ancestors' powers of abstract reasoning were much less developed than their survival skills.

It is difficult to imagine how they might have thought. The modern human mind has benefited from many thousands of years of social grooming. Trying to imagine not being able to read, say, is doomed to failure since even the briefest glance at a document will usually give some clues. Studies were made a few decades ago, however, of contemporary primitive peoples, hunter-gatherers whose way of life was probably similar to that of the artists of Altamira. Peoples such as the Andaman Islanders in the Indian Ocean and the Bahau Dayaks of Borneo are consummately expert at survival, but it was found that they had difficulty in recognizing abstract concepts. Their system of counting was—one, two, many—some primitive languages having no number greater than two. Small stones are used to help handle larger numbers. If such primitive procedures might encourage a sense of superiority, we should recall that the English word *calculate* has a Latin root that means "pebble." Except for periods of a year or two, contemporary primitives seemed to have no conception of the past. Their more remote experiences were said to have happened "once upon a time" or "a long time ago." Distance was measured in terms of a day's journey, and for measuring quantity, they used such words as "a handful," "a heap," or "a basket."

With so few abstract skills, the relationship between cause and effect would have been difficult to recognize. Moreover, they would have tended to give credit for an invention to the benevolent action of supernatural beings. Thus, novelty would be associated with magic, and therefore outside direct control, a tendency that might not be totally alien to tech-

*See Paul Ehrlich (2000) for an excellent review of our early development, and separation from the Neanderthals.

nophobes today. However, our ancestors were exposed to a multitude of misunderstood natural phenomena. Coupled with the mishaps of everyday life that could inflict such unrelenting pain and suffering, it would mean that every substantial advance would be very hard won. Seminal developments such as the bow and arrow or the cultivation of farinaceous plants seem to have been the result of many decades or centuries of interminable trial and frustrating error. As ever, they would be driven by the courage, audacity, or brilliance of a tiny few who stubbornly refused to accept things in the form that Nature had provided them.

Eventually, our ancestors' indefatigable persistence culminated in the ability to produce and manipulate iron. This may not seem much to write home about today, but Lewis Morgan (1877), writing in the nineteenth-century, described this moment as "the event of events" in human experience, a discovery without parallel or equal. It gave us the hammer, the axe, the anvil, the iron-tipped plough, and the sword. See Text Box 3. He made these remarks during a time of supreme mechanical inventiveness, of course, but even from today's broader perspective, maybe he is right. There is no question that the acquisition of iron technology transformed our prospects and horizons.

Remarkable though this event was, however, the discovery by the Greeks at roughly the same time (ca. 1500 B.C.) of an efficient and flexible way of writing language alphabetically (rather than through pictures and symbols) was at least of comparable importance. Pens thereafter could be mightier than swords. It opened a dimension of humanity's intellectual

---

Text Box 3: Iron and Welding

The ancient Greeks too would seem to have recognized the momentousness of the discovery of iron. Herodotus (1996) in his wonderful *Histories* writes about King Alyattes sending presents (in approximately 620 B.C.) to the Oracle at Delphi as a tribute in return for the recovery of his health. He gave a large silver bowl and a salver of welded iron, of which the latter was judged "the most remarkable of all the offerings at Delphi" (p. 11). He said the salver was the work of Glaucos of Chios, the inventor of the art of welding. Some of the offerings to the Oracle were remarkable indeed by present-day standards, and for comparison had included "a vast number of vessels of gold ... weighing in all nearly 2,500 lb" (p. 8)—a tribute that would be worth (roughly) about $20 million today. Welding at that time would be by hammering. It was not until the nineteenth century that the modern art of welding was invented.

Text Box 4: Foundations of Civilization

In his book *Ancient Society*, Lewis H. Morgan (1877) compared humanity's achievements up to the time when we could still be regarded as living in a primitive state with what we had done afterward. It is a remarkable comparison. Even if our civilization were reduced to rubble, there would still be an infinite gulf between the way of life of the survivors and that of the animals. As he puts it (pp. 34–35):

[Humanity's] ... achievements as a barbarian ... transcend, in relative importance, all his subsequent works. Starting as humanity did of course at the bottom of the scale, those achievements up to the time of the iron age included poetry; mythology; temple architecture; knowledge of cereals; cities compassed by walls of stone, with battlements towers and gates; use of marble in architecture; ship building with planks and probably with nails; the wagon and the chariot; metallic-plate armour; iron sword; wine; potters' wheel; mill for grinding grain; woven fabrics; axe and spade; hammer, anvil and bellows; ... plus a large variety of legal and social achievements, including marriage and the family.

development whose potential is probably unlimited. The Greeks were not the first to make this vital breakthrough. The Hebrews and the Phoenicians, among others, had done so centuries before. But the Greeks used it to launch an unprecedented period of intellectual expansion, many of the fruits of which we still enjoy today.

Morgan's event-of-events comment should not be underestimated. Humanity's progress had been miraculous even up to the times we condescendingly describe nowadays in terms of Stone, Bronze, or Iron Ages. Such prosaic references give no hint of the towering intellectual, technological, and social achievements made by our ancestors during the relatively short time that separated them from the primates. See Text Box 4. Indeed, they tend to obscure them. The magnificent cave paintings of Ardeche and Lascaux, for example, were in archaeological terms the works of Stone Age man. Homer, who lived around 800 B.C., was, with similar accuracy, an early Iron Age poet. These descriptions imply that their achievements were naive or primitive, but who today would not envy their skill?

Homer was one of the most influential, imaginative, sensitive, exciting, profound, perceptive, and inspiring writers who ever lived. Furthermore, his genius was not isolated. It is now widely accepted that Homer may not himself have created the stories told in the *Iliad* and the *Odyssey*. He seems to have gathered them together from traditional sources—of which perhaps the performance of easily remembered rhyming ballads was the most important—as few people could write. Homer wrote about the triumphs and frustrations of the great, the good, and the ordinary; about their relationships with the immortal gods; and about the fantastical schemes and intrigues the gods perpetrated on each other and mortal humanity. Perhaps his greatest contribution was the idea that the written word could endow a type of immortality. It is one thing to hope that your exploits will be remembered in ballads long after you are gone. It is another to see the words recorded on durable parchment that will actually be spoken or sung, especially if you can write them yourself.

Homer was one of the first of a small number of remarkable people who came along in the following few centuries who were to transform our way of life. The most influential was Aristotle some 500 years later (384–322 B.C.). Isaac Newton once said that if he had seen further than others, it was because he had stood on the shoulders of giants. This was certainly true of Aristotle too. The foundations for his encyclopedic work had been laid by masters such as Sophocles, Euripides, Socrates, and his pupil and successor, Plato—people whose genius would grace any age. The Greeks recognized that such problems as the essence and structure of civilization, law and punishment, philosophy, politics, religion, and the meaning of life were subjects that could be debated and analyzed. They were not fixed for us. They made up the fabric of society, and we could weave them as we wished.

Aristotle made good use of his elevated stature. See Text Box 5. He was the first to write about thinking as a process that could be subject to disciplined and rigorous examination. He also seems to have been the first scientist of any description—natural, social, or political. He was indeed an intellectual superman, and it is difficult to overestimate his contribution. Imagine a fantastical person with the combined intellects of, say, Darwin, Descartes, Kant, Marx, Popper, and one or two other giants according to taste who with some further magic could be transported back in time to ancient Greece and educated according to the fashion of those days. Aristotle would appear indeed to have had all the abilities of such a fanciful and fictitious person. Aristotle lived well over 2000 years ago, but go to any decent research library today, and you will find shelf after shelf of works written or inspired by this amazing person.

Aristotle was lucky in his choice of patrons. Philip of Macedonia ap-

Text Box 5: Aristotle

Only a small proportion of Aristotle's original writings has survived, and for many of these we are indebted to his dedicated student and successor, Theophrastus. In addition, the written material has deteriorated over time and no doubt has been reconstituted and copied many times. Nevertheless, some 2000 printed pages have survived. Apart from the Scriptures, they form the most enduring legacy of any writer in history.

pointed Aristotle to tutor his 13-year-old son, Alexander, a post he held for five years. Alexander was strongly influenced by Aristotle's teachings, in particular by his faith in the intellect. But, of course, his ambitions extended to a much larger domain. While still in his twenties, Alexander the Great,* as Aristotle's prodigious pupil became, conquered much of the then known world and created the world's first superpower. Not surprisingly, therefore, Aristotle lacked for nothing. He was the world's first well-funded scientist and was free to do as he pleased. It seems he had the equivalent in today's money of some £10 million[†] at his disposal every year, from which riches he was able to set up a large institution—the Lyceum—and support the research of several hundred workers at home and abroad.

A little later, one of Alexander's generals, Ptolemy, founded a dynasty in Egypt. History is not exactly full of examples of conquerors taking pains to foster the intellectual well-being of the vanquished. But perhaps as a reflection of the Greeks' supreme confidence in the future, Ptolemy honored his leader by setting up a university at Alexandria—the Museum. Its library would have graced any great city today and attracted numerous researchers from Eastern as well as Western cultures. They included such geniuses as Euclid and Archimedes and the less well-known Aristarchos of Samos. The latter was the first to propose the theory that

---

*Scientific advisers appointed by Aristotle accompanied his army. Alexander was probably the richest man in the world, but it was said that his most treasured possession was Homer's *Iliad*, which accompanied him wherever he went.
[†]This is a very rough estimate. Aristotle's income for the Lyceum was said to be hundreds of talents a year. Aubrey de Selincourt, in his translation of Herodotus's *Histories*, says that a drachma was the daily wage in the fifth century B.C. for a skilled worker in the Athenian navy—that is more than a century before Aristotle—and 1 talent was equal to 6000 drachmas. My guess is based on current British Navy rates of pay for able seamen.

the sun was the center of the universe. It did not go down well, however, as Aristarchos's proposition would have demoted the earthly divinities, and he was lucky to escape unscathed. Thanks to Archimedes, his proposal was not forgotten, but it had to wait for Copernicus in the sixteenth century before it received the consideration it deserved.

Aristotle was strictly an intellectual. Transport him to the proverbial desert island, and he would not have survived for long. This would be true of many people today, but they would tend to regard it as a weakness. In Aristotle's case, however, he would consider it a strength. He believed that educated people should not labor—that was for slaves and other menials. Their task was to understand the minds of the gods and to try to emulate them. The word *Museum* had a different meaning then. Literally, it meant a place connected with the Muses, the presiding geniuses of the literary, musical, and scientific arts. Thus, one visited the Museum to worship the Muses by studying them. It would have been out of place to possess any practical expertise (other than writing) because it would have a corrupting effect. Archimedes, for example, was famous in his lifetime for his contributions to such down-to-earth fields as mechanics and hydraulics. Nevertheless, Plutarch, the Greek literary genius of the first century A.D., wrote of him:

> He looked upon the work of the engineer and every art that ministers to the needs of life as ignoble and vulgar.

Archimedes was strictly a theoretician.

From today's perspective, it is almost incomprehensible that the Greeks did little to improve their material well-being.* Aristotle studied Nature and particularly the "living creatures," but his most enduring legacy was that *thinking*[†] about a problem was an end in itself. On the social scene, the Greeks laid down the legal and democratic foundations

---

*M. I. Finlay (1963), in his book *The Ancient Greeks*, writes: "Workmanship of the finest quality was abundant. Good craftsmen were constantly improving their knowledge of materials and processes, in ways that left no trace in the written records. Nevertheless, the fact remains that the basic Greek technology was fixed early in the archaic period (from 800 to 500 B.C.), both in agriculture and manufacture, and there were few major breakthroughs thereafter. The list of Greek inventions is a very short one indeed" (pp. 127–128).

[†]Plato, for example, was bitterly opposed to empiricism. He said that if there was ever to be progress in astronomy, the actual appearances of the starry heavens must be disregarded. They should be seen to be what they were and not taken for absolute realities.

> ## Text Box 6: Innovation
>
> An abhorrence of innovation has not been restricted to the Greeks or the Romans. Christian teaching ensured that it was largely the norm up to the time of the Renaissance in the West, and some environmental evangelists would urge its return today. The Islamic East has been similarly hostile. Lawrence James, reviewing *Lords of the Horizons* by Jason Goodwin in *The Times* (London) of March 26, 1998, quoted from Islamic scripture: "Every novelty is an innovation, every innovation is an error, every error leads to Hellfire." He went on to relate that when at the close of the eighteenth century a Russian fleet appeared in the Mediterranean, (Ottoman) officials were genuinely puzzled as to how it had got there and concluded that since it had not passed through the Bosphorus, it must have been carried overland. The idea that there might be other routes simply did not occur to them.

of civilized life. No intellectual issue was ignored. The creative spark was therefore fully ignited, and that is often the precursor to substantial economic and social progress. It did not materialize. Although the Greeks' dissent drove powerful engines, they were, in modern language, without gearing or transmission; that is, they were not connected with the tangible world of the everyday. See Text Box 6. Their respect for intellectual purity was virtually absolute. As a result, the peoples that could build such architectural miracles as the Parthenon made no further contributions to the technological infrastructure that had made such constructions possible. They were among the best rule-of-thumb engineers the world has known. Paradoxically, therefore, intellectual brilliance was accompanied by technological stagnation. The Greek civilization then began a slow descent perhaps because of its inability to innovate or to recognize that innovation was important.

The Hellenistic Age reached its greatest brilliance in the century following Alexander's death (323 B.C.). The outpouring of creative talent was vast and covered every aspect of life from the serious to the scandalous and salacious. There were said to be more than 400,000 books (largely written on papyrus rolls) in the Alexandrian library alone. See Text Box 7. The end of that century was marked by the rise of Rome, whose legions rapidly conquered the territories annexed by Alexander and much more besides. Remarkably, however, the fierce and relentless Romans were captivated by the Greeks' accomplishments and lost little time in

Text Box 7: The Library at Alexandria

The tragic loss of one of humanity's finest libraries—the Library at Alexandria—has long been subject to debate. Up to the time of Edward Gibbon's (1796) definitive *The Decline and Fall of the Roman Empire*, it was accepted, according to a historian writing in the thirteenth century (Abulpharagius), that the Muslims destroyed it in A.D. 642. He relates that an appeal was made to Caliph Omar for the library to be saved, and he famously replied with one of the earliest examples of the twentieth century's Catch 22: "If these writings of the Greeks agree with the book of God they are useless, and need not be preserved; if they disagree, they are pernicious, and ought to be destroyed" (Vol. III, p. 568). The rolls of parchment were then supposed to have been distributed among the 4000 baths of the city for heating purposes, whence they lasted six months. The story is not credible. If parchment has, say, 20% of the heating value of good coal, (about 8 megawatts per ton), it is doubtful if even 400,000 rolls of parchment (say, a kilogram or so each) would have kept so many baths hot for more than a week or so. The story would seem to have amusement value only. As with other losses from ancient times, the "specifics," in modern military jargon, have not surprisingly been lost. However, we should be grateful, as Gibbon recommends, that so many writings from that enlightened era have been preserved for us despite the ravages of accident and malice.

assimilating the parts they liked best. Latin, for example, was then crude and imprecise. Within a few generations, however, Greek scholars had polished the language of their conquerors into the sophisticated and literary tongue we know today.

The abstract and theoretical, however, had little appeal to the hardboiled Romans. The Hellenistic sciences were largely ignored, therefore, except for those with an impact on agriculture. Education was encouraged. In the western part of the Empire, the language of instruction was naturally Latin, but in the East (and in the Arab world in particular) they wisely chose Greek and so preserved their contact with abstract traditions. Unfortunately, few people could speak both languages. Consequently, a cultural gap developed which progressively detached the Roman leadership from the theoretical insight that could inspire change.

The Romans' emphasis on education ensured that expertise would be quickly diffused throughout the Empire. Trade expanded and the capa-

Text Box 8: The Romans and Research

Apparently, the Romans did not even have a word for the modern concept of experiment. Their word *experimentum* meant trial or tribulation, a usage that comes from Hermes Trismegistos, the priest, prophet and sage, and possible contemporary of Moses. Trismegistos thought that Nature's secrets could only be extracted through torture by fire, distillation, or other chemical manipulation. The reward for success would be eternal life and youth as well as freedom from want and disease. The Romans were also fatalists, having three goddesses of fate or destiny—Nona, Decuma, and Morta—the Fates. If something went wrong, the Romans blamed it on opposition from the Fates. If something went well, they attributed the happy outcome to the Fates' cooperation. With such an outlook, the Romans had no need of research!

bilities of each province were gradually brought up to the standards of the best. But innovation was strongly discouraged, and so there were no new technologies or new ways of doing things. As the provinces of the far-flung Empire assimilated each other's skills, therefore, the rates of production converged toward the maximum sustainable yields. Once achieved, the total volume of trade could expand thereafter only if the number of people involved in this zero-sum game could be increased. See Text Box 8. But population did not grow. Indeed, it fell, and there was no further territorial conquest. Economic stagnation is bad enough, but to make matters worse, the arrogant Roman plutocracy could not curb its greed. Its demands became more extortionate, and the inevitable consequence was that productivity fell as the Empire's subjects became ever more disaffected. The Roman economy began to sink into decline from about the second century after Christ.

The reasons for the fall in population are still not wholly understood.* However, no doubt an increase in disease against which they had no remedies, the incessant calls for fighting men to defend the Empire, and the debilitating demands for taxes imposed by an increasingly cor-

---

*Present-day Russia may be having a similar experience. *The Economist* reported in April 2000 that Russia's population had dropped by 6 million over the past decade. It quoted an estimate from American demographer Murray Feshbach, who believes that worsening health and continuing poverty may reduce Russia's population from the current level of 143 million to as little as 80 million by 2050.

rupt, ruthless, bloated, and blundering bureaucracy on peoples systematically starved of hope would hardly have boosted fertility.* Furthermore, many parents murdered their children rather than condemn them to misery. Religious fervor often thrives in these circumstances. Increasingly, people turned to the new religious movement that was open to all and which offered salvation in the next life—namely Christianity—the rise of which mirrored the gradual fall of Rome. The full story is vastly complicated, of course, and studies of the decline would fill a library of Alexandrian proportions. The enduring fact is its speed. In only a few centuries, the spirit of the Greco-Roman civilization to which we owe so much was gradually crushed in the West, although thankfully its seeds were preserved for posterity in the more prescient East.

Perhaps as an amateur historian, I might conclude this portion of my back-of-the-envelope history with another word from that consummate professional, Edward Gibbon. In his *The Decline and Fall of the Roman Empire* first published in 1796, he takes the reign of the emperor Philip (A.D. 244–249) as marking the beginning of the end. He writes (Vol. I, p. 155):

> To the undiscerning eye of the vulgar, Philip appeared a monarch no less powerful than Hadrian or Augustus had formerly been. The form was still the same but the animating health and vigor were fled. The industry of the people was discouraged and exhausted by a long series of oppression. The discipline of the legions which alone, after the extinction of every other virtue, had propped the greatness of the state, was corrupted by the ambition, or relaxed by the weakness, of the emperors. The strength of the frontiers, which had always consisted in arms rather than fortifications, was insensibly undermined; and the fairest provinces were left exposed to the rapaciousness or ambition of the barbarians, who soon discovered the decline of the Roman empire.

---

*Chester G. Starr (1983, p. 676), writing about the fourth century A.D., said, "The Empire was sinking into barbarism long before the barbarians came."

3

The Rise from Oblivion

The gloom of the Dark and Middle Ages* had finally descended. Expressions of individuality were severely discouraged, and one's fate was determined at birth. But humanity's indomitable spirit continued to flicker and flare. Some flares were brilliant, most notably from such supreme intellectuals as Bede and Alcuin in eighth-century England. They could have sparked a revival in less turbulent times, but the Vikings, among others, had no more respect for intellects than for life or property. Not surprisingly, there was no recorded research, and humanity's material capabilities remained fossilized at much the same levels as at Aristotle's time. The first rumblings of revival came from an international community of wandering friars. During the twelfth century their intense debates

---

*My principal focus in this book is on Western civilization. There have been others, of course, and our lives have been enriched by their legacies. The Chinese in particular had a structured society long before the West and made many important scientific and technological advances. They did not ignite an explosion of growth as they eventually did in the West, however. The reasons why global economic development was led predominantly by Western civilizations are elegantly summarized by Jared Diamond (1997).

Pioneering Research: A Risk Worth Taking, By Donald W. Braben
ISBN 0-471-48852-6   © 2004 John Wiley & Sons, Inc.

led to the creation of universities at such enlightened places as Bologna, Oxford,* and Paris. This was a vital step in humanity's rise as it nurtured the idea that secular learning had a value apart from theology. These institutions began to break the monasteries' monopoly, but it was a long time before they escaped religious control.

One of Rome's most enduring legacies was Latin, which became the universal language of scholarship throughout the West. In principle, ideas could therefore be freely exchanged across frontiers. Unfortunately, they could just as easily be centrally controlled. The language of Western Christianity was originally Greek, but after the second century it was succeeded by Latin. This change had nothing to do with religion. It simply came about because Rome was the superpower at the time. However, the change also created a major problem that took centuries to resolve. Religious issues dominated all scholarship then. "Classical" studies, however, that is, those requiring knowledge of Greek, were regarded with intense suspicion in the Christian West because of the association with paganism practiced by the Greeks for centuries. Thus, Christianity inadvertently created a barrier separating Western scholars from the achievements of earlier civilizations. Books on these advances had been preserved in the East, but they had, of course, been originally written in Greek.

A powerful stimulant for change came from the awful Crusades— that two-centuries-long series of brutal and merciless wars prosecuted in the name of religion. Remorseless and malicious intent has rarely led to more unforeseen benefits. Millions died, but the Crusades eventually broke the aristocracy's iron grip on an effectively enslaved people. This was simply because so many members of the upper class—the driving force behind the Crusades—lost their lives. Another fruitful peace dividend stemmed from the renewal of contact between the intellectually starved West and the Eastern provinces that had generally remained cultured after the fall of Rome. Adelard of Bath, for example, traveling in the disguise of a Muslim student, brought back the first rudiments of physical and mathematical science from the schools of Cordoba and Baghdad. The Greek–Latin language barrier slowly began to crumble.

The echoes from a more cultured past were later to inspire Roger Bacon (1220–1292). His masterpieces summarized the wisdom of the day and

*Keith Feiling (1966, p. 154) gives a vivid description of university life in early Oxford: "where two or three thousand boys, between fourteen and twenty-one years of age, chose their own masters, ruled their own hostels, roamed the streets with bows and arrows to attack townsmen and Jews, sank to animal depths in the taverns, soared to the highest themes of philosophy and salvation."

attempted to promote science to its proper status in university curricula—all European universities were then, either directly or indirectly, subject to papal control.* Bacon's truly revolutionary step, however, was to recognize that thinking alone was not enough. Aristotle himself had always recognized the importance of observation, but that idea had died with him. As a result, his astonishing but unintended bequest, apparently immutable for over 2000 years, was that conclusions based on rigorous logical reasoning *were the equivalent of facts.* The religious establishment drew much of its strength from his unimpeachable authority.† Unfortunately, Bacon's heroic bid turned out to be unsuccessful. He did not give up even after being imprisoned for his audacity. Together with others largely forgotten, he remained tireless in his efforts to liberate philosophy from the minutiae of medieval theological debate, and played a vital role in preparing minds for the scientific revolutions to come.

Fame is often short-lived, but a remnant of William of Occam's imprint on history has survived the passage of time. Occam (or Ockham, 1285–1349) was a courageous scourge of fourteenth-century woolly thinking, and he too has long been one of my heroes. Born in the former village of Ockham in Surrey, he was an ardent student of logic. He is chiefly remembered today for his extensive use of the saying *"non sunt multiplicanda entia praeter necessitatem,"* or "entities are not to be multiplied unnecessarily." It may be translated into more modern English as "do not presume more than you have to." Perhaps Aristotle had inspired him, as he had always maintained that Nature "did nothing in vain." Thus, Occam insisted that all observations must eventually yield to explanation, which in turn should be as simple as possible. Indeed, the learning process is a matter of successively removing redundant assumptions about causes and effects as we gradually approach the kernel of truth. It is somewhat surprising that Occam's pedantic prose has penetrated the centuries. Its demand for succinctness, however, must have infuriated the autocracy. Occam made such devastating use of it in his fearless demolition of their waffle and obfuscation that it came to be known as Occam's razor. We should remember that it has not lost its edge.

The dawn of revival optimistically heralded by Bacon, Occam, and other dauntless dissidents proved to be false. Their ideas failed to catch on not because they lacked in quality or clarity. There are none so deaf

---

*Nowadays, a mischievous person might remark that while we no longer have papal control, universities are often controlled by pervasive and inflexible bureaucracies with similar aspirations to infallibility.
†Medieval scholars used the phrase *magister dixit* ("the master has spoken") to denote an irrefutable argument, thereby invoking Aristotle.

---

as those who will not hear, of course, but it is difficult for minds to focus on intellectual issues when disaster threatens and day-to-day survival is a serious problem. The Great Plague of 1347 to 1351—the Black Death—and its recurrent ravages throughout the following 50 years and more, was perhaps the biggest catastrophe to befall any civilization in war or peace. The carnage of twentieth-century war and repression does not come close to matching it. Estimates vary, but it seems that it swept away about a third of the population of England and similar proportions elsewhere in Europe and the Near East. We now know that the cause was the bacterium *Yersinia pestis** transmitted by the fleas of infected rats. But at that time, the plague was utterly incomprehensible to professional and peasant alike. Ignorance does not stop conjecture, of course, and pseudo-explanations abounded. Dissidents are at extreme risk at times like this as people who raise their heads above the parapet can easily lose them. After some 80,000 had died in Paris in 1348 to 1349, it was thought that the end of the world was coming. King Philip VI believed that the plague was God's punishment for their sins and ordered that blasphemers be punished by forfeiting their offending lips. Apparently, the sins of the time were not believed to include war: Despite the plague, England and France continued their bitter hostilities for another hundred years.

Not surprisingly, the plague was also followed by intense and prolonged social unrest. The area of cultivated land decreased for lack of laborers, and the peasants struggled constantly "to put hunger to sleep" as William Langland—the gaunt poet of the poor—put it in his contemporary poem *Piers the Ploughman*. Remarkably, for a time when change occurred slowly, if at all, the cost of labor more than doubled in little more than a year. The survivors suddenly woke up to the fact that they could take their talents where they pleased. The feudal system was already under severe stress from the loss of aristocrats during the Crusades, and this new development accelerated its demise. The Black Death proved to have other welcome side effects for those lucky enough to escape its clutch. The paralyzing grip of the guilds and of the apprenticeship laws was eventually broken, thereby increasing mobility. It also created a new class of entrepreneurs who seized the opportunities for trade and profit as population began slowly to rise again. Never has a pestilential wind blown so much good.

The dawn of revival finally broke at about the beginning of the six-

---

*The discovery was made by two scientists—Alexandre Yersin and Shibasaburo Kitasato—working independently at the same laboratory in Hong Kong in 1894. The bacterium was originally called *Pasteurella pestis*.

teenth century. One of the key catalysts seems to have been the discovery of economic ways of making paper. The Arabs had known about it for centuries, and they had learned it from the Chinese. The returning Crusaders brought it West—yet another priceless peace dividend. At about the same time, Gutenberg in Mainz and others such as Caxton at Westminster were developing the techniques of typography and printing. These discoveries transformed the intellectual climate. Before about 1450, books were handwritten on vellum, a fine parchment made from calfskins. The number of books circulating in Europe was measured in thousands—that is, about 1% of the number kept at the Alexandrian library alone some sixteen centuries earlier! By 1500, however, it seems to have increased to about 5 million, and books were slowly broadcasting news of the dawn among a receptive populace that had long been starved of intellectual stimulation.

Civilized societies are like great incubators. If the environment is dominated by a rigid respect for authority and tradition, the spirit of adventure is quickly quenched. Challenges to what is thought or done are suppressed, and ignorance prevails. Such societies do not change—indeed they stagnate. Conversely, when there is education, freedom of movement and information, and a cultural environment that tolerates originality and innovative thinking, there is sustained and healthy social and economic growth. We have been lucky that something approaching these Renaissance conditions have usually been the norm over the past several hundred years. For some periods, and for some regions, there have been and continue to be severe lapses, of course. The environment will perhaps never be ideal for everyone. However, individual productivity has increased more than a hundred-fold since the Renaissance. As human ingenuity seems to be limitless, material prosperity should continue to rise as long as conditions on our great incubator are generally favorable.

Sustained growth necessarily takes us into uncharted waters, however, and we cannot expect that progress will be either painless or predictable. In the sixteenth century, the newly acquired printing presses spread learning far and wide. They also began to fan the flames of critical debate and to stimulate serious scrutiny of ways of life accepted unquestioningly for centuries. Religion was, of course, a key issue. But it was not so much the Church's faith that Martin Luther (1483–1546) challenged as the extravagances and abuses of its principal administrators, particularly those who lived in Rome. Luther's famous 95 Theses nailed to a church door in Wittenburg in 1517 would probably have passed unnoticed a few decades earlier. See Text Box 9. But the thunder of his defiance was amplified by the prolific presses and struck resonant chords first in northern Europe and later much wider afield. The reverberations from what began as a

Text Box 9: The 95 Theses

The 95 Theses were a closely argued set of statements attacking the Pope for the sale of indulgences—the scandalous practice by which rich people were persuaded to part with vast sums of money for the papal coffers against his guarantees that their sins would thereby be forgiven. Reading an English translation of them today, after the passage of almost 500 years, one is struck by their reasonableness, as the first six might indicate:

1. When our Lord and Master Jesus Christ said, "Repent" he willed the entire life of believers to be one of repentance.

2. This word cannot be understood as referring to the sacrament of penance, that is, confession and satisfaction, as administered by the clergy.

3. Yet it does not mean solely inner repentance; such inner repentance is worthless unless it produces various outward mortification of the flesh.

4. The penalty of sin remains as long as the hatred of self (that is, true inner repentance), namely till our entrance into the kingdom of heaven.

5. The pope neither desires nor is able to remit any penalties except those imposed by his own authority or that of the canons.

6. The pope cannot remit any guilt, except by declaring and showing that it has been remitted by God; or, to be sure, by remitting guilt in cases reserved to his judgment. If his right to grant remission in these cases were disregarded, the guilt would certainly remain unforgiven.

But the Pope's word then was law. To criticize him was blasphemy, and as a result, Luther was excommunicated. Luther, however, was a very passionate man who knew how to convey his scathing invective to the common people in the earthy language they would understand—"the Pope shits lies" was one of his typically offensive remarks. His response to excommunication was to burn the Papal Bull of Excommunication in public. It is miraculous that he escaped unscathed. He not only escaped, of course, but also inspired a revolution. In sharp contrast, John Wyclif (1330–1384), the English protestant, was burned as a heretic (posthumously!) merely for translating the Scriptures into English.

simple protest led to the Catholic Church's greatest upheaval since its formation more than a thousand years earlier.

One of the unforeseen effects of the ensuing Protestant Reformation was to extend elementary education beyond the classical bounds of Latin and Greek and into the vernacular. This was not what Luther had intended. He was said to be appalled by the very thought of reason and toleration. He had sought to replace the Roman system by a more austere one of his own but with precisely the same claims of infallibility. The Protestant Church became, however, "a priesthood of all believers" rather than one of an elite and privileged few. Thus, the long-standing monopoly of the clergy in mediating spiritual transactions between God and humanity was ended. Every Protestant now had an *active* role to play. A lay "calling" or vocation came to be regarded as being just as important as a clerical one. Whatever Luther's intention, the increasing emphasis on vocation was precisely what was needed. Individual creativity could at last be expressed, thereby stimulating commercial activity and technical innovation. The Europeans finally began to move away from medieval mediocrity.

Scientists too began to make progress. Some began to take the hitherto unthinkable step of questioning Aristotelian philosophy—a scientific form of blasphemy. Aristotle thought that every natural phenomenon could be explained by the fact that the four elements of which everything was comprised—earth, air, fire, and water—struggled to attain their ultimate place in the divine order of things. Thus, according to convention, the reason heavy bodies fell to the ground was because Earth was the center of the universe and hence the ground was the natural resting place of everything above it. As has been mentioned, the Greeks were largely untroubled by the need to test their theories by experiment. It was enough to ensure that their ideas had a defendable logical structure and were self-consistent. Inspired perhaps by Roger Bacon, scientists now became much less dogmatic than the confident Greeks. They began to see that research could yield practical applications and help people live easier and healthier lives. Youthful science had at last struck out on its long and illustrious career.

It had to serve a long apprenticeship, however. Contributions from such giants as Copernicus, Kepler, Galileo, Descartes, and Newton were mainly cultural and inspirational. It is not that scientists had no wish to see practical application of their work. Society did not yet turn naturally to them for solutions to practical problems. Technical progress had always been sporadic. Each lurch had usually come from trial and error and the determined efforts of anonymous people utterly convinced of the superiority of their own creations. Youthful science now began to play a steadily increasing role, but for some time innovation continued to be led by expediency. We should not forget, however, that other universal stimulant—the prospect of rich rewards.

The word *technology* seems to have come into use about 1615 and is derived from the Greek *technologia*—the systematic treatment of an art. Progress has always depended on the social climate. In particular, people will strive to make better mousetraps if society acknowledges successful inventors and can protect their just rewards from being stolen. The concept of ownership is, of course, as old as the hills. It now came to be associated, however, with the *ideas* essential for the replication of a piece of hardware or of a manufacturing process, as if the ideas themselves were tangible objects. This novel interpretation of an old concept came to play a vital role in stimulating another revolution.

Sovereigns in most European countries had long been able to confer rights on persons or organizations to operate monopolies.* The documents giving these rights were necessarily addressed to the public, and so it had to be possible to "open" them without breaking the sovereign's seal. Hence, they were called letters "patent" from the Latin for opened or exposed. Not surprisingly, these rights were routinely and blatantly abused, but perhaps because of its long established democratic traditions, the scandals eventually raised the strongest storms in England. See Text Box 10. The opposition was so powerful that Parliament was eventually forced to act, and these corrupt practices were declared unlawful by an act of Parliament—The Statute of Monopolies—in 1624. The statute contained a very important exception, however. Exclusive rights to exploit a *new invention* could be granted to an inventor for a term of 14 years. This simple measure soon came to play a substantial role in the creation of a new superpower.

Alfred Whitehead, the English mathematician and philosopher, has aptly called the seventeenth century the "century of genius." It was, of course, a European phenomenon, but Francis Bacon (1561–1626)—later to be Lord Verulam and Viscount St. Albans—was one of its most influential figures. He is also one of the most complex and controversial characters in scientific history. Trained as a lawyer, he became a successful barrister, entered Parliament, and eventually, through a series of astute alliances and wheeler-dealings that were exceptional even for a time when intrigue was commonplace (the Gunpowder Plot, for example, was in 1605), won the highest office in the land—that of Chancellor to King James I of England (James VI of Scotland).

---

*The sale of patents for licensing "orderly" inns and alehouses, for example, which were anything but orderly.

Text Box 10: Magna Carta

At Runnymede on the River Thames between Staines and Windsor, the *Magna Carta* (Great Charter) was discussed by King John and his rebellious barons, agreed to, and signed in a single day—June 15, 1215. It was not so much an agreement as a set of enforced mutual concessions. Each party recognized that agreement was better than the endless hostilities of the lack of it. The king furiously complained that they had given him "five-and-twenty over kings," as a council was to be set up comprising 25 barons appointed to enforce the orders of the charter on the king. But the barons were forced to concede the advantages of putting some voluntary limits on their powers. The charter had many imperfections. The rights of the peasantry (the villeins) were excluded, for example, as the charter discussed the rights of "freemen" only. However, it was the vital first step toward the establishment of a form of parliamentary democracy eventually adopted the world over. If one day in a millennium should be celebrated, it surely should be this. The charter conferred the right that "no freeman shall be seized or imprisoned, or dispossessed, or outlawed, or in any way brought to ruin ... save by legal judgement of his peers or by the law of the land"—that is, not merely because his sovereign or a baron orders it.

Bacon had a very ambitious agenda for England's role on the European scene. Unfortunately, politics then as now usually involves compromise. Bacon could only make progress if he could find ways of working with his patron, the treacherous Duke of Buckingham, and with the king. As ever, there was a price to pay, and he had to enter into a most miserable compliance with corruption at the highest levels. It should be mentioned, however, that corruption was endemic at that time. No official was paid at anything like the rate needed to maintain a position. Bribery was a universal necessity, therefore. Nevertheless, his period of office was described as "the most disgraceful years of a disgraceful reign." He was involved, among many other things, in corruption through the sale of peerages and offices of state and in the sale of monopolies* against which

*Clayton Roberts (1966, p. 29) discusses "the disgust at the sordid patents that Bacon had certified" and the astonishment at his carelessness in accepting gifts. However, A. L. Rowse (1950, p. 377) points out that Bacon was more sinned against than sinning. The bribes had not affected his judgement. Indeed, "he had given sentence against the givers."

the Commons was later to act so decisively. Although it was customary then for chancellors, and indeed all office holders, to receive gifts or gratuities from grateful suitors, it put him in a difficult position when he meekly tried to protest against the royal excesses. Even this lukewarm, half-hearted action was enough to lead to his downfall. Buckingham disowned him, which was fatal as the Duke's favor was essential. Bacon was caught in the bitter crossfire between the Commons and the king and was impeached in 1621, being declared unfit to hold office or to sit in Parliament.

All this might make one suspicious of Bacon's accomplishments. That would be a great pity. One of the shrewdest and most sensitive commentators of the time was Ben Jonson (1572–1637). Known as honest Ben, he was widely regarded as second in importance as a dramatist only to Shakespeare himself. Jonson wrote of Bacon:

> My conceit of his person was never increased towards him by his place or honors. But I have and do reverence him for his greatness that was only proper to himself, in that he seemed to me ever by his work one of the greatest men, and most worthy of admiration, that had been in many ages. In his adversity I ever prayed that God would give him strength; for greatness he could not want.

What a wonderful accolade! And this in spite of the fact that Bacon had recently been convicted of corruption and sleaze!

Throughout this turmoil, however, Bacon also maintained a passionate interest in science and philosophy and published extensively. See Text Box 11. His works were widely acclaimed throughout Europe, but he was sometimes as capable of misjudgment in science as he had been in politics. He scorned Copernicus's work, for example, which in turn won him the contempt of his scientific contemporaries. William Harvey (1578–1637), who discovered the circulation of the blood and was the physician to James I, once sarcastically remarked that "The Lord Chancellor wrote on science like a Lord Chancellor." Bacon's apparent lapse should not be regarded as too surprising. Copernicus's theory of the universe was still intensely controversial. As the Vatican had tried quietly to point out to Galileo before it instigated his trial, Copernicus's ideas were still a hypothesis. English universities, for example, forbade their teaching until the 1640s. Bacon detested "armchair theorizing" and would have been very suspicious of any claims not founded on the extensive and disciplined accumulation of experimental data.

His impeachment having freed him from the distractions of high of-

Text Box 11: Accademia dei Lincei

Similar ideas to Bacon's were simmering in the land we now call Italy and in the mind of a passionate advocate for science in particular, the eighteen-year-old nobleman Federico Cesi. In 1603 he founded the world's first scientific society—the Accademia dei Lincei (Academy of the Lynx)—an international society determined to observe the natural world with the keenest possible eyesight, hence the reference to the lynx, rather than rely on what Aristotle and others imagined it to be. Galileo was to become its most famous member. Unfortunately, Cesi died suddenly in 1630, and in the face of the intense opposition from the Church and Cesi's own family, the society died with him. Pope Pius IX revived it in 1817. It lost much of its independence during the Fascist government but emerged at the end of World War II to continue as one of the world's most respected institutes.

fice, Bacon could now employ his eloquence energetically* to advocate the value of science. He did so with gusto. During his last few years— "the spent hour-glass of my life"—as he described them, he accomplished more of lasting value than he ever did in an office of State. In his *Novum Organum* (published in 1620) he developed the arguments to justify why conclusions should be based exclusively on scientific data rather than from a system of explanations laid down by "ancient philosophy" (i.e., Aristotle). In particular, he stressed the power and importance of research, the relationship between "systematic knowledge and technics," and the ways that the advancement of knowledge will lead to new discoveries that are beyond our present imagination.

It will be recalled that the Protestant Revolution established a priesthood of all believers, and not merely of academic specialists. Similarly, Bacon promoted the idea that science and technology should be open to anyone with talent. In his *New Atlantis*, he proposed that scientific research should be state supported. Scientists had a duty to learn from the

---

*His writings became increasingly influential after his death. Their quality and elegance were such that many, particularly in the nineteenth century, thought he was the real author of the works of William Shakespeare. This continues to be an intriguing question, but the case for Bacon is very weak.

ways of craftsmen* and thereby "to draw out of them things of use and practice for man's life and knowledge." Thus, he was among the first to see clearly the potential of the changes in thinking taking place around him. He had not only the courage and conviction to express them but also the genius to compose them in language that would endure. Bacon had therefore sown the slowly germinating seeds of the Industrial Revolution. He was[†] to inspire many followers, and a small group of them went on to found the Royal Society in 1660.

In the hope that a few present-day dissenters might also be inspired to take us beyond our imaginations, I have selected the following sample of Bacon's admirable philosophy, Bacon (1620), which should of course be corrected for the lack of symmetry in its treatment of the sexes that prevailed in his day (the italics are mine):

> It will perhaps be well to distinguish three species and degrees of ambition. First that of men who are anxious to enlarge their own power in their country, which is a vulgar and degenerate kind; next that of men who strive to enlarge the power and empire of their country over mankind, which is more dignified but not less covetous. But if one were to renew and enlarge the power and empire of mankind in general over the universe such ambition (if it may be so termed) is both more sound and more noble than the other two. Now the empire of man is founded on the arts and sciences *alone for Nature is only to be commanded by obeying her.*

I shall return in Chapter 7 to the modern implications of Bacon's intriguing use of the word "obeying." Many people today seem not to subscribe to this Baconian view, believing, apparently, that Nature can be commanded by consensus or sufficient money.

At the beginning of the eighteenth century, the climate of innovation was most favorable in Britain,[‡] although France and Holland were not far behind. Large deposits of easily mined coal and iron ore gave Britain

---

*He was an exceptionally clear thinker. He recognized, for example, that the humble workshops of metalworkers, glass makers, paper makers, dyers, brewers, and sugar refiners were in effect the "laboratories" of the time.

[†]He was a scientist to the end. While studying the effect of cold in preventing animal putrefaction, he stopped his coach to stuff a fowl with snow. He caught a fever from which he died.

[‡]A. L. Rowse (1950) argues that the rise of industrialism stretches back to the middle of the sixteenth century and to the freedom allowed to individuals by Queen Elizabeth and her ministers. Indeed, it was not until the late eighteenth century that the rate of industrial expansion was again as high as it had been in her seminal reign.

a head start, and the exploitation of these resources in turn stimulated the need for more efficient transport—canals at first but rapidly followed by roads, ports, and better ships. All this needed investment, and there was a similar expansion of the banking world—400 new banks were started in England between 1780 and 1815. Astoundingly, in spite of the enormous demand for money, interest rates were low and capital was cheap, thereby encouraging entrepreneurs to take risks and plan more enterprises. British governments had generally been noninterventionist since the brilliant Elizabeth's reign. Corners could usually be cut with impunity and profits taken at whatever level the market would bear. Furthermore, labor costs were low, a fact that was ruthlessly and systematically exploited. But the Industrial Revolution, as it came to be called, was built on technology. One might summarize the reasons why Britain led the way as follows:

- The ground had been very carefully prepared since Elizabeth's time.

- The financial, legal,* and political infrastructure encouraged innovation.

- Britain had a strong Protestant culture that emphasized education[†] and the role of the individual.

- That culture created a spirit of enterprise and regarded the pursuit of profit as an acceptable way of life.

- Dissenters were not actively persecuted and so innovation could flourish.

- Not the least, British inventors knew that their "intellectual property," that is, the patents that recorded their personal stake in their inventions and the income they expected them to create, could readily be protected.

British technology also led the world in most sectors, and its people then were simply more resourceful and inventive than others. However, it had no monopoly on insight. From our modern perspective, we can now see that searching for technology in fields where there is little or no understanding of Nature's wonderful ways is like walking with one's eyes shut. Some light began to penetrate as the growing fund of scientific

---

*England was the only European country in which commercial and maritime law simply became a branch of the ordinary law of the land.
[†]Since the reign of Elizabeth I, Britain had been unique in its demand for scientific literature written in the vernacular.

knowledge began to offer clues on the best paths to follow. Inevitably, however, it was a long time before this option was routinely available. The concept of research as a systematic discipline had to be developed. The sources of technology could then be expanded progressively from an exclusive dependence on the inspired guesswork of isolated individuals to organized searches for solutions to specific problems. Within a few decades, the Industrial Revolution had spread rapidly throughout Europe and North America. Its steady growth has continued—with occasional explosive surges driven by the imperatives of war—up to the present day. I did not intend, however, to write a history of science and technology. My purpose has been to outline some of the factors influencing the ebb and flow of intellectual development since our escape from the animal kingdom. Having launched, so to speak, the Industrial Revolution, we can now let it take its course.

4

Taming Research:
The Problems of Success

It might be amusing to speculate on the life we might have enjoyed if Carl Linnaeus's name for our species—*Homo sapiens*—had accurately reflected our most important characteristic, as it was supposed to do. The deeply religious Linnaeus believed that the human race was the Creator's ultimate achievement, and so it is not too surprising that he chose *sapiens*, the Latin for "wise" or "knowing." As *H. sapiens*, we might have expected, therefore, that our existence would be one of erudite contentment as we shrewdly surveyed, in the succinct words of Douglas Adams, life, the universe, and everything. We might have sat around in calm contemplation, sharing our enlightened visions with similarly well-informed friends, exchanging and comparing philosophies, and generally passing the time in a relaxed state of broad-minded bovine benevolence. Our little children, if such a tranquil people could fire themselves with enough passion for the processes of procreation, might actually have behaved as they were supposed to, once upon a time, by being seen and not heard as they patiently sat at the feet of their elders slowly soaking up the collective

Pioneering Research: A Risk Worth Taking, By Donald W. Braben
ISBN 0-471-48852-6   © 2004 John Wiley & Sons, Inc.

wisdom. There would be no stress and no disagreements, and warfare—
that supreme expression of human folly—would be impossible because a
species whose every member was wise could not possibly conceive of it.
Perhaps true bovines do indeed enjoy such a serene way of life? Or might
it be that they regard our traditional description of them as stupid and
dull with quiet, detached indifference—yet another indication of how
little we humans understand.

Unfortunately, my suggestion of *Homo dissentiens*, or dissenting man,
is much nearer the mark. Dissidents can be dangerous. They can transform peoples' lives. Indeed, there is an ancient Chinese curse—may you
live in interesting times—that would inflict on its victims the misery that
often follows in the wake of radical change and which the fates would
seem to have hurled at humanity from time to time throughout its history. In this respect, the nineteenth and twentieth centuries could possibly be the most interesting we have had, so far, as almost every decade
brought greater changes to our way of life than those that hitherto en-
compassed centuries or sometimes even millennia.

Our glacial progression came to an end with the Industrial Revolution beginning around 1750 to 1800.* Any rises in regional wealth before
that time were succeeded by roughly equal falls. Net growth was neg-
ligible for century after century. Now humanity was suddenly hurled forward like snowflakes in an avalanche. Steady and sustained expansion
continued inexorably year after year. Economic growth accelerated again
at about the beginning of the twentieth century, but in its latter half, the
pace became even more frenetic. Some of the causes of the Industrial
Revolution were outlined in the last chapter, but why did its tempo suddenly shift a gear or two?

Major changes rarely have simple causes. There is a vast literature,
and the financial, legal, and political whys and wherefores feature prominently, as do the technological. One subtle factor, however, seems to
have gone largely unremarked. Originality in the form of dissent and
a passion for experiment has always played a vital role in forging our
future, but until about the turn of the twentieth century, this trait was
largely wild and uncultivated. The great discoveries that transformed our
lives usually came from prolonged trial and error and sometimes from
flashes of genius. They are not predictable. See Text Box 12. We continued to reap this random harvest until well after the introduction of in-

*Angus Maddison (1995) prefers 1820 as the launching point. Before that time, he
notes that technology was virtually stagnant and evidence for advances in economic well-being is very meager. This work is Maddison's sequel to *The World
Economy in the Twentieth Century* (1989).

Text Box 12: Margaret Thatcher

In May 1971, Mrs. Margaret Thatcher (now Lady Thatcher), then the U.K. Secretary of State for Education and Science, appeared before the Select Committee for Science and Technology.* Her responsibilities included the Science Budget. She was asked whether "fundamental research ought to be geared to a practical application at the earliest opportunity, for exploitation in the world markets outside?" She replied:

> I do not know what sudden demands will come. I am only eternally grateful that someone had the wit to back physics and chemistry research so that we could provide radar when the demand came very quickly at the beginning of the war. I suspect that had that judgement been left to the politicians of the day to adapt all scientific research to urgent immediate requirements we should never have retained that capacity.

dustrial research laboratories. Many senior industrialists often took pride in identifying and protecting their maverick geniuses and took steps to ensure that their creative protégés were excused the day-to-day necessities of making money. Their altruism was also enlightened self-interest, of course. They had confidence in their judgement and knew that their patronage would almost certainly be amply repaid. As I shall try to explain, however, even the most senior executives today must bow to the homogenizing pressures of corporate consensus, and the universities are similarly constrained.

Education is now open to virtually everyone in the industrialized world. The job market covers a wider spectrum than ever before. In principle, therefore, there would seem to be a profusion of routes by which talented and determined people can make an impact no matter how humble their origins might be. In practice, however, the outlook is not so rosy. Maverick members of the anonymous rank and file today may sometimes find it more difficult to get support for their new ideas than their predecessors did in the past. This remark may seem surprising. Many will disagree. They might say that the pace of change and the scope of science and technology have never been greater. It might seem absurd to claim that radical thinking is more likely to be stifled today than it has been for a long time. But the remark brings us to one of the main themes of this

Second Report from the Select Committee on Science and Technology, Session 1970–71, House of Commons, July 1971.

book, and my task now is to add credibility to what some will see as an unsubstantiated assertion.

Until the turn of the twentieth century, research was largely the province of lonely and persistent pioneers or academics in their ivory towers. The frequency of their successes, however, encouraged industrialists to bring together groups of like-minded people to help to create new opportunities for exploitation. A little later, governments began to develop the infrastructure vital to the success of an industrial society. The evolution of reliable means for achieving these ends—that is, the building of laboratories and the development of viable structures and organizations— did not happen overnight, but the story of twentieth-century science and technology is well known. It might be sweepingly summarized, however, as tracing the gradual transition from the days when advances came almost entirely from the dogged determination of inspired individuals to the last few decades of the twentieth century when their efforts had been tamed and brought under strict control.

From a material point of view, the transition brought unparalleled success. In economic terms, the average gross domestic product (GDP) of every person on this planet has increased fourfold in real terms since 1900 even though global population has tripled during that time (Maddison, 1995).* This prodigious rate of growth continued for decades before its origins were accurately diagnosed. Adam Smith (1776, Bk. 4, Ch. 2) said:

> Every individual necessarily labors to render the annual revenue of the society as great as he can ... He intends only his own gain, and he is in this led by an invisible hand to promote an end which was no part of his intention.†

Smith's seminal work was an attack on the rigid laws and practices that worked against the entrepreneurial spirit. He consulted far and wide in Britain and Europe—Hume, Johnson, Voltaire—and Benjamin Franklin offered invaluable advice on the colonies. Smith's book took an international perspective and was very influential among Europe's leadership. Although innumerable economists have addressed the problem of growth over the years, Smith's "invisible hand" of the market continued to fig-

*The above estimates are based on the outputs of 199 nations—virtually the entire global population.
†A similar remark may be made about scientists who are free to explore. Thus, if scientists were as free as they once were, the invisible hand would foster growth whether or not it is the scientist's intention. Indeed, that was precisely what happened during the Golden Age.

TABLE 1. Global Economic Growth

Phase	Period	Global Annual Average Compound Per-Capita Growth Rate
I	1820–1870	0.6%
II	1870–1913	1.3%
III	1913–1950	0.9%
IV	1950–1973	2.9%
V	1973–1992	1.2%

Source: Maddison (1995).

ure prominently. In the twentieth century, it apparently worked overtime. Some 40 years ago, however, Robert Solow* at the Massachusetts Institute of Technology introduced what he called a "new wrinkle" in separating the per-capita contributions to growth that come from capital—finance, land, and resources in general—from those which come from technical change and concluded that technical change is by far the major source of economic growth. This factor was complex, however, and included many elements, such as new knowledge. With breathtaking understatement, economists generally came to call it the "residual," even though at that time it seemed to account for some 90% of growth! Solow's discovery and his extensive work on growth theory won him the Nobel Prize for Economics in 1987.

Global economic growth has not advanced smoothly. Maddison's panoramic work identifies five distinct phases and estimates the real average compound annual growth in GDP per head for each period (see Table 1). In phase I, the United Kingdom was the technological leader, and 63% of world growth took place in Europe. Growth accelerated in 1870, a time that also saw important political changes—the end of slavery in the United States and the emergence of Germany and Italy as modern nation states. Technological leadership passed to the United States in phase II. Phase III was marked by the unprecedented turbulence of two world wars and a financial collapse (1929) of global proportions. The rate of growth fell, but despite these appalling circumstances, the world economy continued to expand. Phase IV has been acclaimed as the "Golden Age." It saw the dismantling of colonialism, the creation of new and influential agencies such as the Organisation for Economic Co-operation

*See, e.g., *The Review of Economics and Statistics*, May 1957, p. 312.

and Development (OECD), the World Bank, and the International Monetary Fund, and the General Agreement on Tariffs and Trade, which involved in Maddison's (1995, p. 34) words "a high degree of articulate economic co-operation, at least among OECD countries." It also produced growth on an unprecedented scale. Phase V saw the great oil shocks of 1973 and 1979 and the full impact of the collapse of the fixed-exchange-rate systems at Bretton Woods in 1971.

Maddison (1989) presents an economist's account of our interesting times in absorbing detail, and his epic story should be required reading in every place of learning. He acknowledges the role that technology has played in all this and concludes (p. 105):

> Throughout this analysis it was assumed that the major engine of growth has been advancing knowledge and technological progress which needs to be embodied in human and physical capital in order to have its impact. There is no reason to suspect that this will change.

However, Maddison's "major engine of growth" is far from being a simple machine. In common with many other human pursuits, science and technology have always been the products of their time. The fact that almost nothing new came out of the Dark Ages, for example, does not imply that people then were incapable of having original ideas. Science and theology were indistinguishable in that dismal era. People were locked into a rigidly hierarchical system that suppressed individuality and punished dissent. The might of authority had also deemed that everything under the sun was precisely as we perceived it to be because it was the expression of God's will. Nothing about it could be questioned or changed. Thus, a period famous for its walled cities was also a time when enquiring minds were similarly enclosed. Since experimentation was equated with sorcery, innovation was severely discouraged in all walks of life for five or six centuries. Indeed, repressive regimes throughout history have censored the works of art, music, and literature they deem to be subversive, and they continue to do so to the present day. Picasso's art was banned in Franco's Spain. Shostakovitch's music was banned for years in Stalin's Russia, and there is scarcely a country in the "civilized" world that has not at some time burned people because of their books. For science and technology, too, the intellectual climate not only strongly influences what can be done but also can affect the quality and potential of these vital sources of economic growth.

Funding is one of the most important factors. From Aristotle's time to the present, patrons have been indispensable to all but the rich for those who wish to pursue their passions professionally. For modern sci-

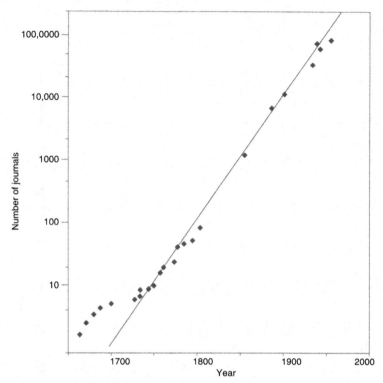

Figure 6
Total number of scientific journals published globally in any one year, which
we take as being roughly in proportion to the number of practicing scientists.
(Reprinted by permission from D. de Solla Price, *Science Since Babylon*, Yale
University Press, New Haven, Connecticut, 1975.)

entists, however, the biggest unsolved problem stems from the spiraling
success of their enterprise. Not surprisingly, every triumph has stimulated
a new wave of recruits, and each in their turn might make the inspirational discoveries that further swell sciences' ranks. And so it goes on.
Statistics is a relatively recent obsession, and data on the variation of the
scientific population with time are not easy to come by. However, Derek
de Solla Price (1975, p. 167) suggests that the number of current scientific
journal titles might give a rough guide.

Almost all scientists are strongly motivated to tell the world about
the great things they have done, and journals successively spawn several
offspring when they perceive that they have more authors than they can
satisfy. Counting journals, therefore, is roughly equivalent to counting

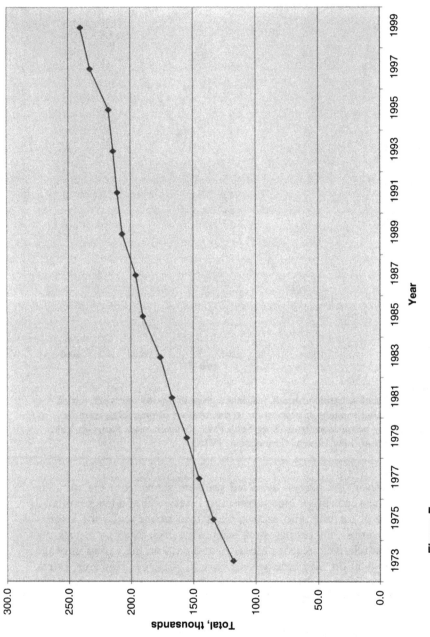

Figure 7
Doctoral scientists and engineers employed in academia in the United States. (Data from

scientists. The earliest known scientific journal—*The Philosophical Transactions of the Royal Society of London*—was first published in 1665 and was soon followed by others in mainland Europe. De Solla Price's estimates of their number up to 1950 are given in Figure 6. The actual numbers of doctoral scientists and engineers employed by colleges and universities in the United States between 1973 and 1999 are given for comparison in Figure 7.

The ascent of man has been remarkable in many ways, but seldom has its progression been so spectacular. Since about 1700, the number of working scientists seems to have grown roughly exponentially, increasing 10-fold every 50 years, or an average of some 4.7% per annum (pa) over the entire period. Figure 7 shows that there has been a continued increase (in the United States) but it has slowed down since 1973 to some 2.4% pa. Population has grown too, but at a much slower rate—a 10-fold rise has taken two-and-a-half centuries. The huge rate of increase in the number of professional scientists most obviously cannot continue indefinitely. Indeed, we may already be approaching the sustainable maximum. The increases have presented the scientific authorities with a formidable problem. How should its growing membership be funded? The general response has been to try to ensure that there is equality of opportunity, but this apparent common sense leads to the problems we shall discuss in the next chapter. Some have suggested the palliative of restricting the numbers of graduates entering research, but any financial benefit would be delayed, as fresh graduates are by far the cheapest scientists to maintain. It would, of course, be irresponsible to encourage young people to go into research if their career prospects were dim. Much more importantly, however, saving at the expense of the young would be rather like eating the seed corn.

The simple fact remains that the funds to keep an increasing army of scientists on the march now greatly exceeds what can reasonably be provided. This funding crisis is affecting the quality of the research being done, as I shall discuss in the next chapter. The implications for technological change will be discussed in Chapters 6 and 7.

5

The Bureaucratic Jungle

Science policy lacks glamor. Few youngsters musing on the future are likely to be attracted by a career devoted to assessing ways in which other people might best spend their time. Who would be a role model? Who might take them aside and offer such kindly advice as "One day, if you work hard enough, you too could be a policy maker"? There would seem to be no profound problems to attract the great minds and no glittering prizes to tempt the bounty hunters. Policies are best judged with the benefit of hindsight, and science policy makers are rarely around long enough to reap either the benefits or the whirlwinds. Furthermore, funding nowadays comes mainly from governments, corporations, or institutions. Policies are usually made by committees, thereby diluting any glamor. There is the dubious advantage, however, that no *one* need feel that he or she should take the blame if things eventually go wrong.

It is a different story for policy makers in other fields. Politicians dominate the everyday world, and while they do not always enjoy the public's full esteem, the allure of politics continues. A political career is an end in itself. It is a socially acceptable alternative to making one's living in, say, industry, the arts, or the law or simply by the sweat of one's brow.

Pioneering Research: A Risk Worth Taking, By Donald W. Braben
ISBN 0-471-48852-6   © 2004 John Wiley & Sons, Inc.

Text Box 13: Patronage

Very few scientists will decline the opportunity to serve on one (or more) of the multitude of committees of national grant-awarding bodies. They would decide who gets funded and who does not. Such bodies can make or break careers, especially if the supplicant is on "soft" money, that is, he or she does not have a tenured position. You get to know the influential people, learn the dos and don'ts of the funding game, and progressively become an influential person oneself. As the time required is modest, these advantages usually make the invitations irresistible.

In science, policy making is a labor of love or duty. See Text Box 13. The scientists called to serve are among the most reputable, but they must squeeze the time from their normal professional schedules, and normally they are unpaid. Unfortunately, scientists tend to be dismissive of colleagues who devote more than a small fraction of their time to matters of research policy or assessment. One would be spared this disdain at the end of a successful career, or for the fixed term of a sabbatical, say, but few scientists in the full flood of their faculties would give high priority to policy issues. This is not surprising, of course. One becomes a scientist because one wishes to advance the understanding of Nature or the world of technology or both. However, there is an additional disincentive for scientists who can read the writing on the wall and feel that they ought to try and change it. It is the fear of being written off as someone looking for a plausible excuse to avoid facing the sterner challenges of the laboratory bench.

Scientists have always had to be concerned for their reputations. Not too long ago, however, it was usually a consideration for the future, especially for young scientists. Time was not necessarily of the essence, and if one chose to do something important or to tackle a difficult problem, then so be it. One would be judged eventually, but only at the appropriate time. Nowadays, few people can afford to take the long view because immediacy of return is paramount. One's reputation *today* is all important, and so young people too are pressured into proving their worth as soon as possible. Contrast this new situation with that expressed by the great American physicist Richard P. Feynman (1989, p. 33):

> We should listen to other people's opinions and take them into account. Then, if they don't make sense and we think they're wrong, then that's that!

Feynman did his wonderful work in the bygone age of a few decades ago. Nowadays, scientists whose vision differs substantially from their colleagues' walk a difficult tightrope. On the one hand, their very professional existence can be dependent on their ability to maintain a reputation as a serious and responsible scientist. Failure to do so might mean that their work might not be funded or their results might not be published in a respected journal. On the other hand, they might feel a duty to test their heretical views. But to get funded nowadays they would have to reveal at least some of their thinking before they can be sure about it. Heretics today must therefore risk undermining their reputations. One might resort to subterfuge or even outright lies about one's intentions, but dissembling may not come easily to the scientist capable of noticing that the Emperor of his or her chosen profession is walking around stark naked.

The all-pervasive funding crisis has arisen only over the past few decades. The university as an institution has survived five or more centuries, and so in principle, today's problems should not present too many difficulties. Universities traditionally were hosts to wide ranges of opinions. Academics were supposed to argue about anything. One might expect that the crisis would prompt solutions galore; therefore, as the various factions proudly proclaimed their own uniquely perceptive prescription. If so, one would be disappointed. Sadly, diversity has been displaced by consensus. It is now widely believed that the panacea for the crisis, and indeed virtually any academic problem, can be found in a procedure called *peer review*.

Although we might not give it such a pedantic name, most of us in any walk of life frequently have to face peer review. It might be a friend's assessment of our cooking, a joke, or our driving or of a politician's speech on the state of the nation. It is an inescapable consequence of community life. Furthermore, it can have considerable advantages. The custom of inviting a fellow expert to express an honest opinion on one's work or ideas at an early stage can provide valuable feedback. It can save us from struggling to reinvent the wheel or from doing something similarly stupid because our impatient enthusiasm has made us miss an obvious point.

However, universities and colleges can sometimes be closed communities. Students come and go, of course, but academic staff may stay for years or decades and get to know each other rather well. Academics are no more sensitive to status than other professionals, but the pecking order among academics tends to be determined by mutual agreement. Not everyone seeks promotion. Many would prefer to be respected for their work rather than their rank. But for the ambitious promotion tends to go to those who can rise above their colleagues in some way that dem-

Text Box 14: Academic Women

One of the reasons for the difficulties of penetrating the glass ceiling for women may stem from the academic tendency for self-government. In 1999 in the United States, only about 12% of senior science faculty positions—associate or full professors—were held by women. It was only 3% for engineering. However, there are signs of change, albeit rather slow. In 1973, women held only 4% of senior science faculty positions, and there were almost none in engineering (National Science Foundation, 1999). In effect, women are subject to double jeopardy. They have to convince their colleagues (men and women) that they are worthy of promotion: They also have to convince them that a senior post should be occupied by a woman.

onstrates their superiority. See Text Box 14. The exceptionally brilliant will normally have no difficulties. For the rest, the ability to weather the occasional drizzle of their colleagues' wit, disdain, praise, or sarcasm is not usually a disadvantage.

Some academics are able to combine scathing criticism with suave complaisance. The so-called Oxford sandwich is a classic example. The recipe is as follows: Take a small portion of polite praise for a colleague on some inconsequential attribute, such as "So-and-so is a good speaker at conferences, isn't he?" Follow it by a few large slices of censure, along the lines of "However, his published work is not very original you know. It's much like the thirteenth chiming of the clock, which makes one doubt everything one's heard before. I'm afraid he'll never get into the Royal." Conclude with something benignly anodyne, like "But one must admire his energy and enthusiasm," which will satisfy your audience on your good-natured urbanity. Oxford has no monopoly on this dish, of course; it will be found with various dressings wherever academics gather to take refreshment.

An academic's pseudocloistered existence, even if nowadays it is largely confined to working hours, has always meant that their peers' opinions cannot be ignored. Other professionals are not so much at risk. Industrialists do not usually care what people think so long as they deliver, and their products sell in the market. Many sportsmen and sportswomen might gamble with their reputations, but they would not be unduly concerned as long as they continue to win on the day.

Until recently, academic mavericks could afford to ignore their colleagues' opinions of what they were doing. They would need to be confident that their way would eventually be proved right, but as long as they were only putting their own careers and reputations at risk, their colleagues' skepticism could usually be disregarded. They no longer have this option. This is because selected groups of their fellow experts now have responsibility for setting priorities and allocating resources. Consensus opinion cannot be ignored.

At first sight, it might seem sensible to manage research in this way. We frequently come across similar arrangements in everyday life. At election time, for example, each political party says what it proposes to do, we vote on these proposals, and the party with the most votes gets to run things its way. Unfortunately, science and democracy are poor bedfellows. They are not completely inimical to each other, however. A democracy, in the sense of a form of government, may foster and protect scientific enterprise and may decide, as Francis Bacon advocated, that it should be financed from public funds. But science is essentially undemocratic. No matter how many agree on the validity of a point of view, a single person with a more viable, accurate, or comprehensive alternative may overthrow it. These lonely individuals may have a tough time in getting their triumphs accepted, but without people like them, we would have neither science nor technology and would still be living in the proverbial caves.

Majority opinion in science has often been wrong. One could fill several books with examples on such once-vital topics as alchemy, phlogiston, and the very existence of atoms or of a "life force." Even well into the twentieth century, the consensus among scientists was that the sun was made predominantly of iron; that DNA was an inert junk molecule that could be ignored; and until about the 1980s that the phenomenon of superconductivity was confined to the metallic conductors. But Nature is indifferent to majority opinion, of course, and awareness of that simple fact can provide that extra incentive to the scientist toiling in the small hours who realizes that he or she might be on the brink of being the first person in the world to make a particular discovery. Nevertheless, most scientists seem to have been persuaded to put that simple fact to one side, subconsciously or otherwise, and marriages between democracy and science have been solemnized everywhere. They are, of course, marriages of convenience. The scientific community tolerates them because it is widely believed that they are the best way forward in the circumstances. Not surprisingly, this union of mutually incompatible partners has been blessed by consensus and caution. To make matters worse, these feeble offspring are now responsible for deciding what research shall be done and who shall do it.

The new arrangements are not always worthless. Indeed, majority opinion in science has turned out to be right more often than wrong. Unfortunately, one has no way of knowing, a priori, what the correct answer is. Nature will eventually confirm our opinion, or not, after it has been "proved" by extensive experimentation. For the vast majority of scientists, however, the primary objective of their research is to refine knowledge in existing fields. Bearing in mind the severe shortage of funds, the most experienced and able people in each field should decide how the limited resources should best be used. Their decisions are not usually controversial, and each discipline's development is managed in the most efficient way that consensus can devise. However, this communal role for peer review is new to scientific enterprise. An expert opinion is one thing; *the consensus of experts is another.* Not surprisingly, these recent extensions to peer review's writ have created a vast bureaucracy. In every country, every discipline, or subdiscipline, or special initiative, or whatever has its controlling committee, and they each have their meticulously defined rules and regulations that must be obeyed to the letter if you want your submission to be considered.

As every individual scientist knows, however, the peer review bureaucracy can only be expected to work in well-established fields. Unfortunately, the longer a field has been established and the more precisely a discipline can be defined,* the less likely it is to lead to anything radically new. Molecular biology, for example, which is now at the forefront of so many advances today, did not exist 25 years ago. Sydney Brenner (1998, p. 1411), one of its pioneering practitioners, said:

> For many years it was widely held that molecular biology was a completely useless subject, a "fundamental" science of no interest to those working on practical matters.

If the bureaucracy's emphasis on priority and relevance that is so common today had been the norm 25 years ago, it is not easy to see how this enormously successful field could have developed. Indeed, *consensual* peer review is inimical to new science. Few major discoveries or inventions are greeted with acclaim. In science, the best in each generation have rarely been quick to recognize what turned out to be the best in the next. That has always been true. Today, however, the all-pervasive peer review bureaucracy is the determinant of excellence, and consequently the natural inclination to oppose major challenges to the status quo has become institutionalized.

*Writing long ago, Irvine Masson (1935, p. 199) said, "a science defined is a science dead, a moored hulk."

Science is a global enterprise, and consensus on the best bets will be the same everywhere. National initiatives in such areas as genome sequencing, AIDS, or warm superconductivity abound. The main variables are the levels of funding. Human ingenuity has infinite variety, but the peer review bureaucracy rarely allows it to flourish. One is not free to choose, say, the positions of the scientific goalposts, their width, or their height or dispense with them altogether. That joyous freedom might also be extended to include the shape or size of the pitch, or the ball, or any other rule or constraint, so that novel scientific objectives might be defined that might advance understanding in novel ways. It would be impossible simply because the more radical one's point of view, the less likely one is to have peers.

The commitment to the peer review bureaucracy has been formalized almost everywhere. For example, in 1993 the British government published the White Paper *Realising Our Potential, a Strategy for Science, Engineering and Technology, Cm 2250.* Such excursions into the scientific arena are rare for U.K. governments, and this was the first official pronouncement on the civil research enterprise for some 20 years. The research councils are the most important sources of support for academic research. One of the White Paper's most important decisions was to give the councils a formal mission to enhance the U.K.'s industrial competitiveness and quality of life.

The new arrangements are now in full swing. The Engineering and Physical Sciences Research Council (EPSRC) is the U.K.'s largest source of academic* research funds. In 1998, an application form for a research grant from the EPSRC ran to seven pages of A4. The actual research proposal was required not to exceed an additional six sides of A4 or use a font size smaller than 10. Scientists were completely free to choose the width of the margins. A diagrammatic "workplan" (maximum one side of A4) should also be provided. Forty-two pages of glossy guidance notes accompanied the form, and applicants had to confirm in writing that they had read them. The other research councils had similar requirements. As of 2003, proposals could be made electronically, but the length limits were largely unchanged. Proposals cannot be submitted unless they are "countersigned" by the head of department and an administrative officer in the scientist's institution. Beneficiaries of the proposed work must be

---

*The research councils also support research at a wide range of institutes and laboratories, which they administer directly. Strictly speaking, they are not part of academia, but much of their work is indistinguishable from academic research.

identified, together with the scientist's ideas on how the results will be disseminated to them. The application forms are truly formidable documents, and, it should be noted, the detailed information they demand is required many months before the intrepid explorers will have taken a single step on their proposed voyage of discovery.

The new rules might be sensible if research objectives are tightly defined. If one has never been to Dublin, it could be useful to know the best and most efficient way of getting there. It would be difficult, however, to conceive of arrangements more likely to inhibit innovation. For research truly worthy of the name, a scientist's vision of the future is usually obscured by a fog of uncertainty that cuts visibility to a few weeks at best. Nevertheless, each short-sighted supplicant must attempt to chart his or her progression over the two to three years of an anticipated award. How would such scientists as Paul Dirac, Dorothy Hodgkin, Brian Josephson, Peter Mitchell, Nevill Mott, Geoffrey Wilkinson, and many others have fared if they were starting out today? These U.K. scientists not only won Nobel Prizes but also made substantial contributions to technology and to wealth creation *that no one could have predicted*. Research today is a global enterprise, and communication has never been more efficient. The well-informed scientist who concentrates on what are generally thought to be the most attractive targets will therefore have many competitors. But some scientists succeeded in the past entirely because they did not follow the madding crowds.

Another important strand of the White Paper was Technology Foresight—an initiative intended to bring industry and academia closer together. The government's rationale included among other things (see paragraph 2.24):

> Programmes which are successful in terms of the quality of research may offer no commensurate economic benefit to the country *if firms and other organisations cannot use the results* [my italics]. The Government believes that a co-operative effort is needed to produce a better match between publicly funded strategic research and the needs of industry and other users of research outputs.

These sentences should have raised a storm of outrage throughout the academic community. Instead, it was greeted by silence. British research has been immensely successful. In the 30 years following 1945, British scientists won some 30 Nobel Prizes in the sciences, compared with 8 in the subsequent 22 years. They also contributed to such profitable discoveries as computers, nuclear power, penicillin antibiotics, monoclonal

Text Box 15: The Birth of the Computer

It is widely believed that ENIAC (Electronic Numerical Integrator and Calculator), developed in the United States in 1945, was the world's first programmable electronic computer. In fact, that accolade should go to Colossus, developed by Alan Turing and colleagues at Bletchley Park some two years earlier as part of Britain's ultrasecret wartime code-breaking operations. Indeed, these activities were so secret that some 25 years were to elapse before those involved were permitted to talk about this British operation. Turing's life was tragic. He was one of Britain's most brilliant scientists, and his wartime contribution was priceless. Nevertheless, shortly after the war, he was found guilty of an offence under the homosexuality laws (which would not be an offence in Britain today) and had to agree to chemical treatment to reduce his urge to offend again. Another factor contributing to his suicide in 1954 seems to have been the failure of British government and industry to give him adequate resources to promote the development of the computer he had done so much to pioneer.

antibodies, radar,* television, and jet engines. See Text Box 15. But with the exception of jet engines, these discoveries initially went largely unrecognized by British industry, a notorious lapse the 1993 White Paper ignored. John Polanyi (1995), a Canadian Nobel Laureate, said:

> the strange belief entertained by Governments that while denying an ability to "pick winners" in industry, they seem happy to attempt the even greater challenge of identifying *which* science will yield the future discoveries to benefit *which* industry.

The Technology Foresight Programme (and its derivatives) is now fully operational. The first step in this monumental exercise was to re-

*As the Cambridge historian Correlli Barnet has pointed out, British scientists had the idea for radar, and Britain was the first to develop the actual operational concept. But (in 1938) there was a bottleneck in the production of thermionic valves owing to British manufacturers lagging behind in research. Britain had to get the valves from America and Philips of Eindhoven.

solve the U.K.'s technological needs into 15 sectors.* Many thousands of academics and industrialists were consulted about the markets and technologies which might emerge over the next 10 to 20 years and were asked to define the research needed to realize them. The exercise *required* hope to triumph over experience. Had the powers-that-be taken similar action after the end of World War II, they might have mobilized academics to achieve such surprise-free objectives as better thermionic valves, more efficient piston engines for aircraft, or new ways of generating energy from coal. Thankfully, they did no such thing, but their wisdom has been set aside. Twenty years ago, the personal computer and the Internet were unknown. No committee could have planned for them. Products envisioned today are unlikely to still be attractive in, say, 2023. Nevertheless, Technology Foresight was set as *the* priority for British science for the foreseeable future.

The program has accomplished a revolution in the way academic research is funded in the United Kingdom, and it has been bloodless. The sectoral "recommendations" produced in 1995 became almost mandatory. In the academic year 1998 to 1999, some 70% of the research supported by the research councils was in areas identified by Technology Foresight. As stated by the EPSRC (1998, pp. 2–3):

> EPSRC holds a positive view of both the principles and the practice of the Foresight programme and will continue to value the output of Foresight as one of the main contributors to policy and strategic development.

In 1997, the program's name was shortened to Foresight. With similar perspicacity, it was also recognized that many of the trends identified in the first round of consultations had become dated, so a second round of Foresight began in April 1999. There were to be 3 thematic and 10 sector panels, each looking at the future for a particular area of the economy. There are no prizes for correctly guessing the life of that later survey, and is it not likely that anyone would bet their own money on it being anything like 10 or 20 years. The current round of Foresight—launched in April 2002—operates through:

*The 15 sectors of the U.K.'s Technology Foresight Programme were Agriculture; Natural Resources and Environment; Chemicals; Communications; Construction; Defence and Aerospace; Energy; Financial Services; Food and Drink; Health and Life Sciences; IT and Electronics; Leisure and Learning; Manufacturing, Production and Business Processes; Materials; Retail and Distribution; and Transport.

Text Box 16: Top-Flight Wisdom from the 1960s

In 1966, the U.K. Council for Scientific Policy, chaired by Sir Harrie Massey and including a glittering galaxy of senior scientists from university and industry, said, in its *Report on Scientific Policy (Cmnd 3007)*, May 1966:

> [T]he nation seems at last genuinely aware that our economic future rests upon advanced technology which itself depends on science for its fundamental concepts. The temptation may have to be resisted to throw all our limited resources into the exploitation of present knowledge, thus cutting back our capacity to advance in the future ... It is necessary at the outset to deal with the misconception that the advance of scientific knowledge can be directed from the centre. This would be to misunderstand the original and spontaneous nature of science. The advance of scientific knowledge cannot solely be achieved by the arbitrary selection of national scientific goals and by committing resources of men and money to them. *Because science is original it is also unpredictable: neither the provenance of a new idea nor its ultimate applications can reliably be foreseen by scientific policy makers.* [My italics.]

This excellent advice should be the foundation on which all industrialized nations build their scientific enterprises. It is timeless. It cannot be faulted. It is tragic that it has been forgotten or ignored.

a fluid, rolling programme that looks at 3 or 4 areas* at any one time. The starting point for a project area is either: a key issue where science holds the promise of solutions; or, an area of cutting edge science where the potential applications and technologies have yet to be considered and articulated.

Thus, British academics are being obliged to climb on the bandwagons of consensus, transient rides that curtail their intellectual freedom and might be terminated at any time. They have joined these fly-by-night crusades without fuss and virtually without controversy. They will not escape so easily. See Text Box 16.

At the end of the twentieth century, there were more than 240,000

*In 2003, the areas were Cognitive Systems, Flood and Coastal Defence, Cyber Trust and Crime Prevention, and Exploiting the Electromagnetic Spectrum.

doctoral scientists and engineers employed in American universities, of which latter there were between 2500 and 5000 depending on the definitions used. In the United Kingdom, there were over a 100 universities and colleges, and some 43,000 academic staff, excluding research students and short-term contract research fellows, were engaged full time in scientific research.* They are huge national resources. Indeed, research today is big business. As many of the people involved are publicly funded, it would be absurd if their efforts were not coordinated in the national interest.

In the industrial sector, coordination usually involves direction. This is a defendable policy as industry must be profitable. However, academic research should be supported for much more complex reasons, and direction should not necessarily be essential. Unfortunately, the powers that control academic research do not seem to realize that unless they consider what they do very carefully, they could easily kill the goose that has laid so many golden eggs. Indeed, it is extraordinary that the dangers inherent in these wholesale changes are rarely discussed anywhere. One of the most insidious modern trends is that science and technology are being treated as if, like night and day, they had a fixed relationship. Science today is increasingly seen as the servant of technology, and research is usually justified in terms of its specific goals. Researchers have not stormed in outrage at the loss of diversity either. Indeed, it has been shrugged off as an inevitable consequence of our straitened times. Nowadays, however, there is a dearth of new sciences, and our increasing technological demands must be drawn from a virtually static scientific pool. Our basic scientific research is being increasingly constrained. The relationship is tightening, therefore, because those responsible for the governance of science and technology are *forcing* them ever closer together. When the distinction between science and technology finally does disappear, the unexpected can never happen. Outputs can then indeed be treated as if they were commodities like steel or video cameras and managed according to market demands. But academic researchers are supposed to be expanding our horizons, generating new options, and inspiring the young. The coordination of that effort, like herding cats, ought to reflect its subtleties. It cannot be enforced.

The research-as-a-commodity alternative has been chosen in almost every industrialized country because market-driven research is easier to coordinate. Remember, however, that we are discussing academic research. Is it sensible to assume that the imposition of milestones, goals,

---

*There are also some 10,500 people working in the U.K. research councils' institutes and laboratories.

and timetables will not inhibit creativity? How should such hungry souls as the Einsteins respond? In the United Kingdom, Foresight ignores these issues and implicitly assumes that academics can be turned into short-term pseudocontractors without adverse effects. Furthermore, industry has no obligation to take up their output. Academics can make an important contribution to the economy but they cannot do it alone. Unfortunately, British industries' relatively low investment in science and technology has long been a serious problem. The 1993 White Paper *Realising Our Potential* ignored it, thereby implying that Britain's failure to capitalize on its successful academic research has been the fault of the universities.

The problems are not confined to Britain. H. Eugene Stanley and his colleagues (1999) at the University of Boston studied the growth dynamics of university research in the United States, England, and Canada. They found that they are quantitatively similar to those of business firms. They conclude their paper by saying (p. 437):

> One possible explanation is that peer review, together with government direction, may lead to an outcome similar to that induced by market forces, where the analogue of peer-review quality control may be consumer evaluation, and the analogue of government direction may be product regulation ... *there seems to be no need to make academic research more like a business—it already is.* Some may claim that the business sector can be regarded as a yardstick for organising academic research. If so, the research departments already behave like business units and hence are sufficiently "effective." On the other hand, others may maintain that the "economisation" of academic research has been pushed too far, and that the research system will become "ineffective" if this continues. [My italics.]

The radical changes brought about by *Realising Our Potential* were endorsed and extended by the ensuing government White Paper of July 2000—*Excellence and Opportunity: A Science and Innovation Policy for the 21st Century, Cm 4814.* The Foresight program is confirmed as a "central driver," but Britain's industrial limitations have at last been officially recognized. The 2000 White Paper says (p. 30):

> As well as universities reaching out and transferring their knowledge to business, we need more companies to use science and technology to create competitive advantage. Our record on this is still very weak. Too many of our companies still lack an awareness of the need to change, or the ability to do so. In many industries we invest less in research and development than our competitors, and our companies are less ready to change and innovate than others. This is particularly true of large firms.

It goes on to say that the government is introducing measures to stimulate investment in enterprise and innovation and to encourage risk taking. This extraordinary official admission of what has long been a British weakness is to be welcomed. But what can government really do to rectify this apparently entrenched problem? The industrial and commercial sectors determine their own strategies. The universities, on the other hand, are virtually defenseless against government intervention, and a renowned British strength is to be further weakened. The 2000 White Paper makes an extraordinary statement (p. 27):

> The universities will be at the heart of this effort to build the knowledge economy. Universities can play a central role as dynamos of growth. *But they will only fulfill that mission if they match excellence in research and teaching with innovation and imagination in commercialising research. To do that they will need the skills and the infrastructure to translate science into products, services and marketable commodities.* [My italics.]

In other words, if British industrial research is inadequate, the universities must step into the breach.

Academic research is supposed to probe and test traditional thinking and values. An unprofitable business soon ceases to be a business. Partnerships between academics and business people can be very successful if they are sensitively managed. If the disparate objectives of these two communities are understood, the best of both worlds can be preserved. Some academics have considerable business acumen, but in my experience they chose the academic life because of the freedom it once offered; they did not want to be business people. It is unrealistic to expect academics to be do-it-yourself marketers.

However, a rigorous study of the Foresight saga and other bureaucratic epics is likely to induce terminal boredom. It can also never be complete as the arrangements are continually changing. I will therefore err on the side of incompleteness by mentioning only one more specific example of the reforms being perpetrated in the name of efficiency.

In 1997, the U.K. Medical Research Council (MRC) announced "the most significant changes for over 30 years" in its support for university-based research. The MRC's aim is now "to develop and nurture productive research and training environments." The council also decided to phase out its "stand-alone project grants" by which individuals or small groups were funded and to make awards to collaborations between interdisciplinary teams. The announcement emphasized MRC's wish to encourage "innovation and risk-taking," but only if the scientists can combine into "Centers or Co-operative Groups." One's environment can play an important role, of course. But *environments* never discovered anything. *People* do that. MRC's record in supporting talented individuals doing their own thing was once exemplary. Henceforth, they will have to convince their local center or group committee that their ideas should be pursued. But committees are hardly ever innovative. It might be worth repeating Sir Bernard Cocks's (a former clerk of the British House of Commons) definition of committees as "cul-de-sacs down which ideas are lured, and then quietly strangled."

Despite all this regulation, I have found that committee members do what they can to reward innovation and to reduce the suffocating impact of the rules imposed on them. However, they cannot ignore the fact that in the United Kingdom each research council is now burdened by a formal "mission statement" urging them to:

> meet the needs of its user communities (industry, commerce, government, service sector) thereby enhancing the U.K.'s competitiveness, and the quality of life. [This is the wording of part of the EPSRC's mission. Those of other councils differ slightly.]

Their freedom of maneuver is severely curtailed, therefore. In the not-too-distant past councils merely had to ensure that they supported high-quality research of "timeliness and promise" as it used to be called. We shall see in Chapter 6 some of the benefits that flowed from this very simple formula.

Scientists who do not want to march to fashionable tunes have serious problems, therefore. Until recently, British nonconformists could usually find help from Britain's famous "dual-support system." See Text Box 17. The universities were to be found on one side of the infamous "binary divide," while the polytechnic colleges, which specialized in technical and vocational training, were to be found on the other. Colleges were free to apply to the councils for research funding, but their wrong-side-of-the-tracks status meant that they were rarely successful. The binary divide was in effect a chasm. The two-tier system ended in 1993 when the Higher Education Funding Council for England (HEFCE) and similar councils for Scotland or Wales assumed responsibility for all tertiary education and the 50 or so polytechnic colleges were accorded the status and titles of universities. But the funding councils also acquired another responsibility. They had to build up the universities' capabilities in areas prioritized by research councils, government departments, and industry. An important advantage of dual support in the past was that universities could often support modest initiatives from their own funds. However, the tight financial controls now imposed on the funding councils have effectively closed this small loophole for creative dissidents. It is not so much that

> **Text Box 17: Evolution of Britain's Dual-Support System**
>
> As the number of universities increased during the nineteenth century, the British government freed them from the obligation to conform to the ecclesiastic orthodoxy of the state church. (University College London was the exception. It was founded in 1826 to challenge discrimination and became the first university to welcome all people—regardless of class, race, religion, or sex.) The universities generally opened their doors to Catholics and Jews, for example, and the government began to recognize the need to provide centrally for university funding. At that time, British universities were autonomous corporations chartered by Parliament or the Crown. They were free to set their own statutes subject to the approval of the Privy Council and were generally free to manage their internal affairs as they thought fit. In 1919, the University Grants Committee was set up to provide the central funding (for England and Wales), but the government did not change their legal status. Universities could dispose of "block grants" as they pleased. Indeed, they could (and usually did) resist attempts at governmental control through, for example, the avoidance of "earmarked grants" which stipulated that funds had to be used for specific purposes. The dual-support system evolved from that time. Until recently, it meant that the research councils supported the marginal costs of research—the equipment and contracted staff—while the infrastructure—the universities, their laboratories, and the tenured staff—was financed in England and Wales by the University Grants Committee and by similar committees in other parts of the United Kingdom.

the universities have lost their discretionary powers; they now have very little discretionary money with which to exercise them.

Another important change was the U.K.'s institution in 1986 of the Research Assessment Exercises. These exercises have been carried out by the funding councils at roughly four-year intervals to help determine the priorities by which departments are funded by them. Thus, not only must every scientist who wants a research grant yield to bureaucratic peer review—that is the same everywhere—but in the United Kingdom his or her *department* must also submit to a similar process. It is a very serious affair. A 5* department wins some four times as much funding per capita from the appropriate funding council as a 3b department. It is bureau-

Text Box 18: Research Assessment—United Kingdom

Grades on research excellence are awarded to university departments if their collective research equates with the following criteria:

5*—Research quality that equates to attainable levels of international excellence in a majority of subareas of activity and attainable levels of national excellence in all others.

5—Research quality that equates to attainable levels of international excellence in some subareas of activity and to attainable levels of excellence in virtually all others.

4—Research quality that equates to attainable levels of national excellence in virtually all subareas of activity, possibly showing some evidence of international excellence or to international level in some and at least national level in a majority.

3a—Research quality that equates to attainable levels of national excellence in a substantial majority of the subareas of activity or to international level in some and to national level on others together comprising a majority.

3b—Research quality that equates to attainable levels of national excellence in the majority of subareas of activity.

2—Research quality that equates to attainable levels of national excellence in up to half the subareas of activity.

1—Research quality that equates to attainable levels of national excellence in none or virtually none of the subareas of activity.

cratic madness. Imagine that an athlete has won a race but an official informs him that his gold medal will be devalued or might not even be awarded at all because he belongs to a club with a low athletics rating! How long would athletes tolerate such a ridiculous regime? Yet these are frequent concerns for researchers belonging to borderline departments.

The fifth in the series of research assessments came in 2001 and graded university departments on a seven-point scale. See Text Box 18. It involved more than 50,000* academic staff in some 2700 departments

*The 1996 exercise was described in the journal *Science* as the world's largest peer-review process.

(including the arts and humanities) at each of the universities and colleges in the United Kingdom.

The exercises virtually paralyze universities for months at a time. Success or failure determines a department's future funding or even whether or not it can continue to be considered as a research department at all. Departments awarded grades 1 and 2 in 1996, lowly marks which went to about 20% of the total, subsequently received no funds from the relevant funding council in support of their research infrastructure. This in effect terminated their collective involvement in research. All was not lost, however. Scientists in deprived departments were not prevented from collaborating with colleagues whose departments had been more bounteously favored, but they could not bring much in the way of a dowry.

Performance in the 2001 exercise was measured by each person's publications (up to four could be offered) "which best reflect the quality of the individual's research work" during the preceding four and a bit years. Departments could choose the staff to be included in the assessment. Politically conscious departments were free, therefore, to exclude those who might be perceived as poor performers and thereby tip the balance away from relegation and a lower rating. They had to take care that exclusions would indeed produce a better overall result because the benefits of an anticipated higher grade would be multiplied by the percentage of staff offered for assessment. These and other details are explained in some 25 pages of the inevitable guidance notes to help departments fine tune their submissions. Assessments were made by the appropriate panel according to the "absolute standards of quality" pertaining to each panel's jurisdiction.* Not surprisingly, the scope for Machiavellian manipulation by astute and corridors-of-power-wise department heads was considerable. Time taken in closely examining the interests of the relevant panel members, for example, would probably not have been wasted.

Recurring phrases in the grading descriptions refer to "attainable levels" of national or international excellence. Imagine you are a scientist. How should you judge whether your research might be of "international" or "national" excellence? What would Michael Faraday have said about his work on electromagnetic induction, say, only four years after its completion, a typical gestation period for an exercise? More importantly, what would his "peers" have said if they had been asked to grade it? Since peers today are supposed to be selected from the ranks of fellow

---

*There were 69 Units of Assessment, the legalistic name given by the funding councils to the subject areas, such as Physics; Chemistry; Earth Sciences; History; and Drama, Dance and Performing Arts. Each had an assessment panel.

specialists, the question would pose something of an awkward problem for an imaginary nineteenth-century funding agency because no one else was working in Faraday's field at the time. Albert Einstein and many other peerless revolutionaries would pose similar problems.

One cannot assume that every scientist might be an Einstein or a Faraday, of course. But research truly worthy of the name should be unique. If it is not, academics at least should not be doing it. This latter consideration does not necessarily apply to industrial research. It sometimes may be essential for a company to go over well-covered ground in the search for better products. In the everyday world, one can be famous for doing something important on the national scene which has also been done elsewhere—building a bridge, say, or launching a successful venture. Academic research, however, is a truly global enterprise: It would be no good to claim that you were the first Briton, say, to make a *scientific* discovery such as a gravitational theory or the structure of DNA unless one were also the first person in the world to do so. *For academics, there are no prizes for coming second.* If your research is unique, as you should hope it is, the qualifying words "international" or "national" have no meaning. Nevertheless, the future of your department may hang on them.

The exercises also mean that ambitious researchers determined to tackle difficult problems using new ideas or techniques are likely to be heading for trouble. This is because the search for funds will probably be fruitless. From a departmental point of view, therefore, the time could have been better spent on safer and more productive endeavors. Pioneers have always had to cope with their colleagues' indifference or hostility. They now also have to endure the implications that their dissident behavior is letting the corporate side down.

The recurring phrases are meaningful only if research is directed to specific goals of agreed-upon importance. Peers abound in these crowded fields, and something approximating accurate measurement can be made of one's efficiency and ingenuity. So, the rules of engagement by which British academics should conduct their battles with Nature have been laid down: You should choose a well-populated field so that your performance can be compared with other scientists' efforts. If you have outclassed a British team, national points would be scored but they may not take your department far up the league. If you have beaten an American team, say, there would be international points for your department, the highest league positions can be reached, and more money would be available for your next battle.

The claim that standards can be measured "absolutely" is, of course, ridiculous. The 1996 exercise led to many examples of glaring inconsistency, but the story of the Lancaster University physics department was one of the most publicized. The exercise gave it a rating of 3a, thereby placing it in the bottom 5% of physics departments. But the widely respected Institute for Scientific Information based in Philadelphia co-incidentally concluded in one of its routine assessments that publications from Lancaster had the second highest "impact factor"—a measure of the number of references made in the global literature—of any physics department in England. Nevertheless, despite this obviously international accolade, the rating stood, as appeals were not allowed. The university had little choice but to impose staff cuts in physics. The faculty were understandably furious but knew they had no alternative but to make the best of it. They did. In 2001, the department was awarded the glittering accolade of a 5* rating, but that was no consolation to those who lost their jobs. See Text Box 19.

In principle, universities may choose to ignore a funding council's rating. However, they can only do so by robbing Peter to pay Paul, as their total allocations are based on the assumption that a council's ratings are accepted, which in general they are. Thus, the proud tradition of academic autonomy extending over centuries has ended in a few meekly acquiescing years. The overall performance of a department is a meaningless concept at best. To reiterate an earlier remark—*departments* do no research; *people* do research. At any particular time, some researchers might be doing well while others might be struggling. But the exercises do not allow for individual excellence (or failure). If your research is

---

Text Box 19: A Very British Assessment

In 2001, I wrote to congratulate Robin Tucker, one of our Venture Researchers at the University of Lancaster physics department, on their newly acquired 5* rating. I said that I was sure he and his departmental colleagues must be very grateful to the "Research Assessment merchants" in stimulating such a vast improvement in the quality of their work. He replied:

> Thanks for your note. I see that your tongue is firmly in both cheeks. Yes we now have to don new clothes before we speak to the emperor! And the emperor will be dazzled by these new clothes and give us all keys that he thinks we deserve. We then have to find the locks for them to fit before the next emperor and his court comes by, seeking even brighter robes.

> It was not always so: "Bliss was it in that dawn to be alive but to be young was very heaven."

highly regarded but you have been careless enough to locate yourself in a low-rated department, you might soon lose your competitive edge. The mathematician Benoit B. Mandelbrot, once an IBM Fellow and now at Yale University, who literally introduced new dimensions to mathematics and the sciences with his discovery of fractals, once said:

> Science would be ruined if (like sports) it were to put competition above everything else, and if it were to clarify the rules of competition by withdrawing entirely into narrowly defined specialities. The rare scholars who are nomads-by-choice are essential to the intellectual welfare of the settled disciplines.

Most scientists would agree, at least in private. Nevertheless, the bureaucratic juggernauts continue to roll.

It is likely that the United Kingdom's 2001 Research Assessment Exercise will be the last. Many think that these exercises should never have been tolerated at all. Unfortunately, these people were not very influential.

The current obsession with efficiency has also virtually disenfranchised scientists without tenure. At one time, research students— "irreverent ragamuffins" as Jacob Bronowski once described them— could reasonably expect, after one or two spells as a postdoc, to join the elite ranks of those free to plan their own research. Nowadays, however, new tenured positions—lectureships, professorships, and so on—are thin on the ground. Many researchers have no choice but to accept successions of short-term appointments, if they can get them. This means that postdocs are constantly plagued by the necessities of making ends meet.

To make matters worse, funding agencies will not usually accept direct submissions from unchaperoned postdocs. They can, of course, take part in research led by their tenured and more experienced colleagues. Indeed, the scientific enterprise would very quickly grind to a halt if postdocs did not enter enthusiastically into this state of enforced altruism as almost every research project relies on the input of at least one of these young hopefuls. At my own institution—University College London— they make up some 40% of the research staff. In the United States they number almost 40,000. They are a huge army, therefore, but I do not see why they should not be allowed to bid for leadership. The justification for this crazy policy would seem to be mainly economic. The number of proposals that must be considered by the hard-pressed agencies is certainly reduced by excluding them, but some postdocs are talented beyond their years, and it makes no sense to deprive them of the opportunity to pursue their own lines of enquiry. As they have proved so often in the past, they are the grit that creates our future pearls. Nevertheless, they must now generally wait until a tenured vacancy comes up. Sadly, most will be disappointed, and so the spark of youthful dissent that once inspired many revolutions has, if not been quenched, been heavily dampened.

I would now like to make a small digression. After some 16 years in nuclear and elementary-particle physics research, I accepted an invitation to join the Cabinet Office in London's Whitehall—the very institution that inspired C. P. Snow to coin the now ubiquitous term "corridors of power." The transition from the relatively simple complexities of a laboratory bench in the north of England to one of the country's most respected institutions was difficult. The Cabinet Office is a place where one risks being broken on the intellectual equivalent of the wheel: Fools are not suffered gladly, if at all. Although very few of my new colleagues had been trained as scientists, I was surprised to find that their lay backgrounds were not a serious impediment. In some respects, their detached viewpoint seemed to give them an advantage. My impression was that they approached their work as a barrister might; that is, they did not seem to care whether they were for the defense or for the prosecution. They were not committed to the outcome, one way or the other. Arguments could be presented, for example, on why Britain should build more nuclear power stations or less of them, and the very best of their abilities would be dedicated to whatever decision they were supposed to justify. I thought at the time that this trait was to be found only among nonscientists, but I was wrong in that too.

To return to my theme: The training of barristers and those who toil in courtrooms generally has been honed over many centuries. Their impartiality plays a vital role in the administration of laws that have also evolved over a similar time. Scientists involved in the administration of their enterprise are the very best, the cream of their cohort, but they have usually received no training in that role. As for the "law" they are collectively expected* to administer, *there is not the slightest evidence that future performance in research can be predicted from a research proposal.* Perhaps such administrators can derive some comfort from the knowledge that their fellows in other industrialized countries have the same problems. Together with others—most notably Rustum Roy of Pennsylvania State University, who has tirelessly toiled against the iniquities of peer review—we have tried to raise this thorny issue in the scientific lit-

---

*The current obsession with committees brings to mind the Roman saying "The Senators are good men, but the Senate is an evil beast."

erature on several occasions, but we rarely get a response. As Roy cogently puts it, no industrial company has its own research peer reviewed by its nearest competitor. Nor would a government defense agency go to its fiercest "competitor" for their opinion. If you might be in a musing vein, could you imagine the following exchange:

> Dear Adolf,
> I have before me a proposal for developing an acoustic system for detecting your submarines. I enclose a copy. Its aims are ambitious, but there are many claims on our resources. I would be grateful, therefore, for an independent opinion from your experts.
> Yours ever,
> Winston.

Alternatively, could you embrace the idea that the chief executive of, say BP, would write to his counterpart at, say, Shell, asking for his opinion on an a far-reaching proposal he had just received from his research director? Such questions are ludicrous, of course. But not in the academic sector, where the present system universally *demands* that similar questions are not only asked but must be answered!

All in all, it is a dismal story. Unfortunately, few scientists are ready to champion radical change because the scale of the intellectual inertia to be overcome is enormous. There is also a more sinister disincentive. Those who rock an overcrowded boat are likely to be the first thrown out. Consequently, so long as funding falls short of requirements, it is widely believed that there is no alternative to the present arrangements. Those who argue otherwise must at best endure the hostility usually reserved for time-wasting cranks. By way of encouragement, however, they might look to Voltaire, the pseudonym of the prolific French author and polymath Francois-Marie Arouet (1694–1778). Voltaire was famous for his satirical mockery of a pretentious and all-powerful establishment and for his passionate belief in the enlightenment of reason. The great and the good flocked from all over Europe to seek his opinion on virtually every subject of importance. He was indeed, as he described himself, "a necessary philosopher in a world of bureaucrats." He was also an astute observer. In one of his *Contes* (stories) written in 1772, he resuscitated an old Italian saying: *Il meglio e l'inimico del bene* ("the best is the enemy of the good"). We would do well to remember this today. Unfortunately, the authorities seem determined to extract the best value for every dollar, pound, mark, and yen spent on research. Are they flying in the face of Voltaire's reason? If you wished to buy say a new refrigerator or a car, you would be foolish not to shop around to get the best value for your money. If you know precisely what you want and can get a discount, you have probably got a bargain. So perhaps Voltaire's maxim is not applicable to the modern world?

Soon after I arrived in Whitehall I had to make a one-to-one presentation to the then Secretary of the Cabinet, Sir John Hunt, who was also the most senior civil servant in Britain. This urbane and formidable man was not a scientist, and our meeting had been arranged to discuss some of the thornier problems of radioactive waste management and disposal. I was supposed to be an expert on this subject and had done my homework for the daunting occasion very carefully. The subject is, of course, highly technical, and so I had prepared what I believed to be an elegant and simple explanation of the difficult concepts. I did not need it. His main concern was with the obvious question of why the problem of radioactive wastes was important—the one question that had not occurred to me! Why indeed should such an Olympian be concerned with rubbish of whatever type? The question shattered my hubris, and I struggled for some moments to regroup my thoughts. It was a chastening experience, but I learned a lesson on the value of simplicity that I hope I will never forget.

So let us persevere with Voltaire's simple aphorism. The word *best* implies choice. Even if we know precisely what we want, we may still dither over whether we have chosen the best holiday destination, meal, or garment from those on offer. Indecision may even cost us our original preference. But we often do not know what we want. One might wish to learn how to play the piano or to acquire some other skill but might not be sure we are up to it. In these circumstances, one might be wary of choosing what is claimed to be the best value for money as cheap lessons can often turn out to be very expensive in the long run. Maybe Voltaire has a point after all? Let us now throw the doors of uncertainty wide open and say you were given responsibility for commissioning academic research. The means to some ends may not only be unknown, they may even be unattainable. Who would seriously believe in these uncertain circumstances that policies based on seeking out the best value for money were viable options? See Text Box 20.

People who always insist on having the best can have depressing lives. They must always be looking over their shoulders in case someone else has something better. Nevertheless, most countries are striving to find the best and most competitive research. Their main difficulty seems to be the identification of the best ways to achieve this anti-Voltairean goal rather than whether or not it is a good idea.

Searches are in full swing. The headline of an editorial in *Nature*, October 17, 1996, said:

Text Box 20: Game Theory

Imagine you are playing at a chess tournament. At, say, move 20, you have an optimal position, and you are now considering move 21. Your move might result in an optimal position for move 21, but you might decide to sacrifice a piece because you can see long-term gain. Good players often resort to tactics which lesser mortals may find incomprehensible, but if they go on to win, the apparently suicidal move that turned out to be crucial will be acclaimed. Expert commentators would probably have disagreed about the quality of the move at the time it was made because they could not possibly know how the player's thinking was developing. On the other hand, games where both players always try to optimize their short-term positions should be expected to end in stalemate. Expert commentators would not be able to fault them on any move. These players cannot take risks, as this is rarely consistent with optimization.

> The systematic judgment of research performance is a growing industry in search of international quality assurance.

This prestigious scientific journal went on to urge the European Science Foundation to "sink its teeth" into the research assessment problem, as in doing so:

> it would provide a much needed service not only for Europe, but for the many other countries in which objective research assessment is increasingly considered as an essential goal.

The "industry" is indeed growing. It is staffed by scientists who should know that the only valid quality assurance in research stems from the accuracy of its predictions. Nature herself is the sole and ultimate arbiter of whether or not the research is on the right lines. The problem is, of course, that the results of Nature's arbitration are rarely revealed immediately, and so *faute de mieux*, we have to endure the bureaucrats' instantaneous assessments.

Many of the changes to research assessment I am advocating could be made by the simple expedient of abolishing the use of the word "best." Best leads to odious comparisons and attempts to compare the quality of research in such dissimilar areas as, say, physics, biology, history, or chalk and cheese that can have no logical basis. Comparisons inevitably lead to competition, but one can only compete if the goals are clearly defined. The best targets, research, bets, or anything else are, of course, expressions of consensus opinion, which, as is well known, can often be wrong. Much more importantly, the criterion of best can only be confirmed with hindsight. It is incredible that this simple word dominates the governance of academic research, which should rather be dominated by expressions of individuality.

In my opinion, the highest accolade one can give a new research proposal is that it could radically change the way we think about something important. A research initiative using that criterion would not need to use the word best because it would be irrelevant. For example, it would be meaningless to try to compare a proposal from Einstein with one from Dirac. It would be clear from the outset that their ideas could transform the world. They would also be unique. Such an initiative should be easy to operate, therefore. Once that single criterion had been satisfied, one would only need to be sure that the scientists had the determination and talent to bring their ideas to fruition. As I shall explain more fully in Chapter 8, Venture Research was indeed such an initiative. Not only was the assessment rigorous, but also the procedures we used were logically defendable. All research that satisfied our simple criteria was supported. That is, we struggled to derive the criteria that would do all our selection for us. There was no need to prioritize. In addition, *thinking* was the key word in our criterion, and thinking is field neutral. Being field neutral, we could escape the suffocating clutches of peer review. Nor did we have to resort to futile attempts to grade research by the geographical extent of its possible influence. There is only one science.

But even if he or she can successfully play the current funding games, the long-suffering scientist has yet another hurdle to overcome before results of completed works can be disseminated. As is very well known, only those efforts that survive the rigorous scrutiny of one's peers are published in the most cited journals. See Text Box 21. The problem is so well known that I will not dwell on it. I will, however, give one example of the difficulties. John Maddox, *Nature's* very long-serving editor, in his valedictory editorial of December 7, 1995 (p. 521), said:

> It is a good joke (which I have often used) that Watson and Crick's paper on the structure of DNA could not be published now. It is only necessary to imagine what people would say if it reached them in the mail; "It's all model-building, just speculation, and such data as they have are not theirs but Rosalind Franklin's!"

It is extraordinary that such a comment can be made about one of the most celebrated papers of modern times. The comment attracted little

> ## Text Box 21: All Scientists Are Equal, But Some Are More Equal Than Others
>
> W. Ford Doolittle (2001, p. 1707), reviewing John L. Dowland's book *The Surprising Archea*, said:
>
> > The older I get, the more I understand that—although there are indeed facts about Nature and we can indeed discover them—the personalities of scientists and the politics of their interactions have enormous impact on the importance we attribute to different facts and on the broadest conceptual frameworks within which we interpret them.
>
> It was a courageous remark about a well-known fact of scientific life. It would be wonderful to see a rebuttal from those who passionately defend the peer-review system.

or no attention. Journals continue to reject papers on spurious grounds. Well-connected people can appeal, but in general, there seems to be no accountability.

I have largely concentrated on the United Kingdom's bureaucratic jungle so far, but it is generally much the same elsewhere, although specific details may vary. The United States dominates the research enterprise and accounts for about half of the global expenditure. The National Science Foundation is one of the world's largest funding organizations. Its responsibilities cover almost all science and engineering except medicine. The Foundation's *Grant Proposal Guide* is a weighty document indeed (it is now available in electronic form). Some 80 pages long, it gives detailed instructions on how research should be described and covers such extraneous issues as grantees' responsibilities for establishing drug-free awareness programs. Conformance with instructions "is required and will be strictly enforced unless a deviation has been approved." Proposals "should be prepared with the care and thoroughness of a paper submitted for publication." It is a tribute to scientists' determination that the foundation annually receives some 40,000 proposals, for which it requests some 170,000 reviews from the community. It is not known how many tons of midnight oil is consumed. The Foundation awards some 10,000 new grants each year.

At a meeting of the American Association for the Advancement of Science, it was once suggested that the colossal effort wasted on unsuccessful grant applications could be avoided. Grants could be awarded by lottery in which every registered scientist could take part. The reasons behind this tongue-in-cheek suggestion should have been taken more seriously. The catalogue of bureaucracy outlined above forces scientists everywhere to spend large fractions of their time on the funding game if they want to succeed. Research is being confined by discipline or relevance to some national objective. Publication in reputable journals tends to be similarly restrained. Thus, scientists are being divided, ruled, and regulated to such an extent that the sciences are being progressively strangled. It is all being done in the name of efficiency.

6

Prospects for Economic Growth

If we can learn about government policy options that have even small effects on the long-term growth rate, then we can contribute much more to improvements in standards of living than has been provided by the entire history of macroeconomic analysis of counter cyclical policy and fine-tuning. Economic growth ... is the part of macroeconomics that really matters.

Barro and Sala-i-Martin (1995)

Britain's industrial revolution spread rapidly around the globe. It also triggered a population explosion that has continued to the present day. World population increased threefold during the twentieth century, but average productivity—the Gross Domestic Product (GDP) per capita—increased in real terms by a factor of 4.3. This unprecedented growth stemmed from technology. But growth cannot be realized unless technology is widely disseminated and used profitably. Without adequate provisions of capital, even a potentially brilliant technology will fail.

Pioneering Research: A Risk Worth Taking, By Donald W. Braben
ISBN 0-471-48852-6   © 2004 John Wiley & Sons, Inc.

Conversely, one of the surest ways to lose a fortune is to put it on a half-baked idea.

My story so far has focused on the factors affecting the technology supply. It is now time to assess how the *quality* of the supply might affect growth in the future. Figure 8 shows the real growth in the global economy between 1950 and 2001. The global average per-capita annual growth per year over the period 1951 to 1974 was 2.8%. The average between 1975 and 2001 was only half that, at 1.4%. Maddison points out that the annual data are subject to considerable uncertainty. They bring together data from some 190 countries on production, employment, countries' exports and imports, exchange rates, reserves, and estimates of GDP at constant prices weighted by purchasing power parity. I assume that although estimates of growth may not be absolute, their ability to indicate trends may be more reliable.

As can be seen, the raw data vary erratically from year to year. They make little sense. The global economy is enormous, of course, and measurements at annual intervals cannot really be expected to reveal real performance. However, such familiar activities as running or walking can be made to look jumpy if they are sampled at the wrong intervals. Movie pictures are perhaps the best example. Human eyes and brains work together to smooth out the sequence of stills projected on the screen and give us a picture that is indistinguishable from reality. If we could not do this, cinemas would not exist. Growth statistics seem to have a similar characteristic, and our senses need a little help in reading their message. If we allow our vision to persist for 5 years, some sense begins to emerge, as can be seen in Figure 9. A more credible picture emerges if we extend it to 10 years, as in Figure 10. Thus, it would seem that the global economy has a response time of about a decade rather than the much shorter periods policy makers seem to use. This is what we should expect, of course. Ten years has often been taken as a characteristic lead time in science and technology. Unfortunately, politicians seem to prefer the attractions of the annual figure's more hectic rhythms, and such jerky tunes as "stop–go," "fits and starts," and "boom–bust" are the consequences of their poor sense of timing.

According to a survey of the world economy published in *The Economist* (September 28, 1996, p. 16):

> Conventional methods of measuring the economy are no longer up to the task, and economists cannot agree on how to improve them. It does not help that within the profession economic statistics is considered a less exciting subject than economic theory, and so it fails to attract as much talent as it needs.

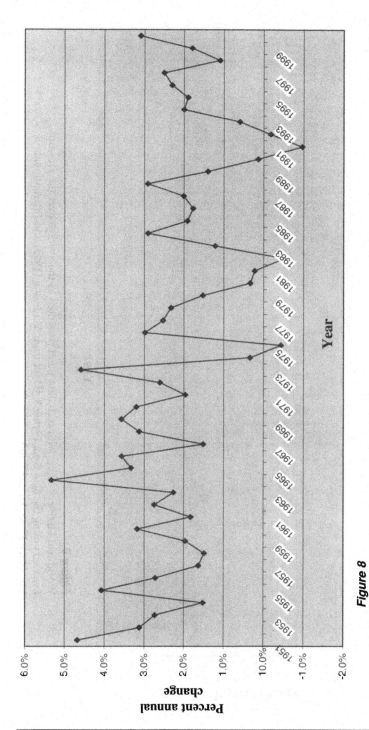

Figure 8

Annual real growth rates of world GDP per capita for 1951–2002 in 1990s dollars. Maddison's (1995) data for 1951–1992 have been extended by those from the International Monetary Fund for 1993–2002, the latest available. World population growth has been taken as 1.6% pa during this latter period.

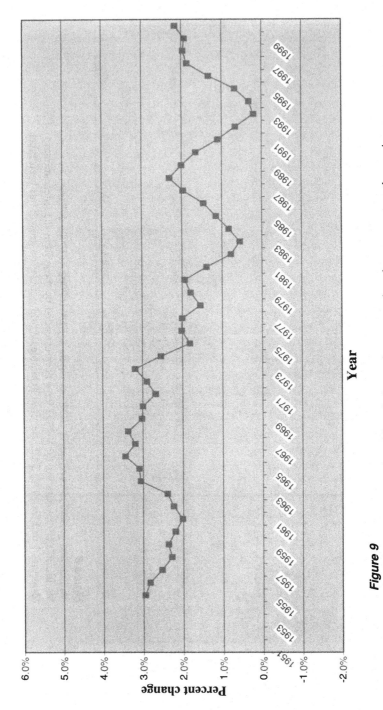

Figure 9

World real GDP per capita with each data point presented as the average over the previous five years using the data from Figure 8. [Data from Maddison (1995) and International Monetary Fund Annual Reports.]

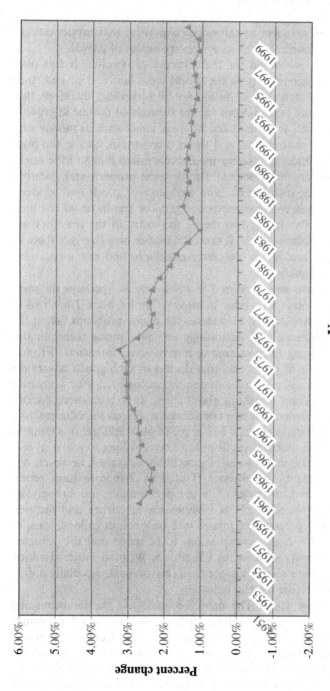

Figure 10
World real GDP per capita with each data point presented as the average over the previous 10 years using the data from Figure 8. [An earlier form of the data presented in Figures 9 and 10 was given in D. W. Braben, *Physica A 314*, 768–773 (2002), and is reproduced by permission of Elsevier.]

The data, however, do seem to influence *confidence*, that mysterious entity which seems to control markets and every aspect of growth.

Economists are renowned for the diversity of opinion. It was once said that if all the economists in the world were laid end to end, they would not reach a conclusion. It should not be surprising, therefore, that economists have many explanations for the substantial decline in growth. Assessments, however, are dominated by *fiscal* issues such as money supplies, exchange rate movements, and the two enormous rises in the price of oil in the 1970s. Indeed, it seems incomprehensible that so little attention is paid to matters technological. People quite properly seek to profit from the production, distribution, and exchange of goods and services. But it seems to be believed that goods and services are universal common commodities. Like the air we breathe or the water in the sea, they are simply *there* and available to all. It does not matter how they got there or precisely what they are. If you are clever enough to find new ways to exploit them, you will profit.

George Soros was described in *The Economist* as "perhaps the most successful financial-market investor in history." In his book *The Crisis of Global Capitalism*, Soros (1998) discusses the major problems facing the global economy. Science and technology are not among them. In this respect, he is following fashion among international financiers. From a scientist's perspective, it seems odd that dealers in such goods or services as steel, pharmaceuticals, or information technologies can be successful without having to know anything about them. It is even more curious that this lack of knowledge is not considered a serious impediment to a financier's ability to make money. For some of us, therefore, it is remarkable that the crisis in global capitalism has taken so long to materialize.

When the fundamental *sources* of growth are discussed, however, one factor is conspicuous by its absence. Technical change features prominently, of course, but estimates of future performance are strongly influenced by the projected levels of investments in science and technology. These overall figures might just as well be given in gallons, tons, or bushels. They readily enable comparisons to be made, but they obscure the creeping uniformity outlined in Chapter 5. With so much attention given to quantity, their *quality* cannot easily be assessed, especially not by lay people. It is therefore usually ignored.

This is a strange lapse. The need for quality in the financing and marketing of a development is well understood, and success often requires genius in these fields too. My favorite example stems from the invention of photocopying. Chester Carlson graduated in physics from the California Institute of Technology in 1930, at a time when even Caltech graduates found work hard to come by. He eventually found a job at the New

York City patent office of a small electronics company (P. R. Mallory Company), but he was obliged to spend a substantial part of his time tediously copying documents. There had to be a better way, and he set out to find it. He was also spurred on by another far-sighted idea—that innovation would help the nation escape from the Depression. It did, of course, but the innovation that revived the economy was also spurred on by war.

Photography was the obvious starting point, but it was expensive and slow. He was not put off, however, and devoted his spare time to research carried out in the only premises available to him—his apartment. In 1938, he discovered a process which he called electrophotography but which later became known as photocopying or xerography. As the latter name implies, it was a dry process and relied on electrostatic attraction between paper and finely ground powder. He was granted the first of many subsequent patents in 1940.

Unfortunately, no one was interested in his great idea. It is always the wrong time for inventors, of course, but having to compete with the priorities of war would not have helped. Eventually, his perseverance was rewarded when the Battelle Memorial Institute, a nonprofit industrial research organization in Columbus, Ohio, agreed in 1944 to develop his copier. His second stroke of luck came a year later when he met Joe Wilson, the president of the Haloid Corporation, a small company specializing in photographic materials. Wilson saw its enormous potential almost immediately. Over the following 14 years, no less, he tenaciously pursued its development despite many setbacks. He was prepared to risk the business, if necessary, to bring it to fruition. Some $75 million was invested—more than his company's total profits for the decade of the 1950s. He was not even deterred when some of the most respected companies in the United States decided, apparently, that the copier did not have a future.

In 1959, the "914 Copier" was ready, but it was some 100 times more expensive than its (highly inferior) competitors. Wilson's sales team, which included Peter McColough, his eventual successor at Xerox, as the company came to be called, were faced with a serious problem. No one wanted their sophisticated machine. Users of the feeble copiers around at the time were apparently content with poor quality, having known nothing else. The team then hit upon a brilliant idea. They would not try to *sell* their expensive copiers: *They would lease them.* Users would pay only for the actual number of photocopies they made.* This stroke of genius

---

*Customers paid $95 a month, the first 2000 copies were free, and thereafter each copy cost four cents. They had the right to return the machines after 15 days, but few did.

led to explosive growth. After nine years, annual sales had reached $1.25 billion, and some 200,000 copiers had been manufactured and placed.

To return to the broader canvas: Economic growth, whatever its source, fluctuates from year to year; world population, however, steadily continues to increase. Its present rate (2003) is some 1.6% pa. While many economists are aware of the need for technological change, they should note that in 1992 the Royal Society of London and the U.S. National Academy of Sciences, in a joint communiqué on population (the first communiqué they had published on any subject), opened with the following remark:

> World population is growing at the unprecedented rate of 100 million people every year, and human activities are producing major changes in the global environment. If current predictions of population growth prove accurate and patterns of human activity on the planet remain unchanged, science and technology may not be able to prevent either irreversible degradation of the environment or continued poverty for much of the world.

This stark Malthusian warning came from two prestigious organizations famed for their circumspection. The communiqué concludes:

> The future of the planet is in the balance. Sustainable development can be achieved, but only if irreversible degradation of the environment can be halted in time. The next 30 years may be crucial.

We seem to be powerless to do much about world population growth for the next few decades. In the longer term, the most desirable way of controlling population is by choice at the personal level. It is probably also the most effective way. Freely available education seems to be the key, a criterion that is only likely to be satisfied when regional social and political climates are ready for it. This happy outcome may be considerably delayed, and so it is possible that within the lifetime of most people alive today world population could double. Before long, therefore, the human race will have to live in a world that will be very different from today's.

Paul Kennedy's (1993) outline of the hazards that may lie ahead is daunting. Social stability is not guaranteed whatever we do, but ignorance will not help. Our survival may depend on the extent to which we can develop our understanding of issues and questions not yet recognized as important. Our best hope of solving this logical nightmare is to encourage and stimulate diversity in everything we do, but especially in science. Unfettered scientific exploration is the primary feedstock of genuinely new technology. It is also perhaps the only way of illuminating unsuspected areas of intellectual darkness. Nowadays, however, exploration is being limited to the fields we know tolerably well. Research enterprise is anything but unfettered. Diversity has been deemed unprofitable.

Short-term survival is vital, of course. Some of the shackles on scientific progress have already been outlined in Chapter 5. As they are not likely to be cast off in the near future, it seems natural to ask from where the new technology necessary to ensure healthy growth will come. Some industrialists say that we already have an abundance of new technology and, indeed, that there are more good ideas than they can develop. But how many of these are step changes, as the transistor, the laser, and the jet engine once were, from what has gone before? Instead, there are a multitude of refinements to existing technologies, where the advantages can often be short-lived as the competition can easily catch up or imitate. Incremental developments rarely lead to the *sustained* growth that breakthrough technology usually brings. These stimuli often came in the past from discoveries made by pioneers who bucked the trends, but such people get short shrift in industry today.

In the academic sector, money for research has always been tight, but until about the 1970s, pioneers did not have to be too precise in disclosing what they were doing. Enough funds could usually be scraped together to test their radical ideas *before* they had to be revealed to the rigorous scrutiny of their peers. Since that time, however, the growing funding shortfall has led to the use of more thorough and businesslike methods to select the research most likely to succeed. Unfortunately, they also tend to throw out the more creative babies with the bureaucratic bath water.

The ubiquitous scientific bureaucracy was created initially by the funding shortfall. It bites with more or less equal severity in every advanced country, although the relative levels of funding may vary. Academic scientists in the United States, for example, have one of the highest levels of funding in the world. British lobbyists would be cock-a-hoop if they succeeded in raising the commitment to British science and especially the status of its scientists to the levels enjoyed in the United States. However, the situation there is far from rosy. When he was President of the U.S. National Academy of Sciences, Frank Press (1988, p. 1) said:

> The United States supports more research than Western Europe and Japan combined, and our system of universities and national, and industrial laboratories is the envy of the world. Why then is our community in an unprecedented state of stress and internal dissension?

In 1991, when he was president-elect to the Board of Directors of the American Association for the Advancement of Science, Leon Lederman said that he would try:

to resolve the apparent paradox of continuing increases in federal research funding and growing dissatisfaction in academic laboratories.

Others have made similar points. But if funding is not the major source of the difficulties, what is? I would suggest that the malaise has its origins in the peer-review bureaucracy. Among other things, it is responsible for the insidious growth of the idea that it is no longer possible to do world-class research without access to the most efficient and usually the most expensive equipment. Laboratories must, of course, be provided with the best equipment. Such resources are essential if scientists are to play their part in reinforcing and consolidating the mainstreams. Furthermore, state-of-the-art equipment, imaginatively used, can open up new horizons if users are allowed to stray from the beaten tracks. But money of itself does not necessarily buy good science. Indeed, many breakthroughs have come from shoestring budgets and the world-class use of the equipment between the ears of our brightest scientists. But the adoption of research policies that concentrate on agreed-upon priorities implies that all the major discoveries have been made. It also ensures that science, like a long-dead language, will eventually lose its vitality and stagnate. Thus, the bureaucracy not only exacerbates the funding problem but also frustrates the efforts of the visionaries who might ease it.

What are the factors that might affect the *quality* of the science-and-technology supply? Government-funded academic research will be taken first. The charities and trusts are also important funders, but they tend to have similar policies to governments. In my view, we have three serious problems.

Serious Problem 1. Science and technology are global enterprises. In science, as I have already remarked, there are no prizes for scientists who come second. In technology, one ignores other peoples' efforts at one's peril, but it is not strictly necessary to be first. Today, communications have never been more efficient. But there is no United Nations, so to speak, for research that might coordinate global scientific policies, nor realistically can there be. As each country determines its own policies and acts in its own national interest, the problem of global coordination is usually ignored. Since each country usually turns to the best-informed scientists for advice on research policies, national policies will be broadly similar. Duplication of objectives on the global scale is guaranteed and the problem tends to be ignored.

Serious Problem 2. A scientist's main objective is to increase understanding. But there are no committees allocating funds in "understanding."

Imagine that by some magic we could be transported to the year 4003, say. It is not certain, of course, that humanity's progression will continue over the next two millennia. We may not survive. But supposing we do, and we overcome such current threats as those arising from population pressure, supplies of water and other essentials, climate change, pollution, epidemics, natural disasters, terrorism, crime, and war. Future horrors presently unimagined should also be added to the list. And we would also have two millennia of scientific progress. Who would not expect levels of future comprehension to be vastly higher than they are now? The inescapable conclusion is, therefore, that the understanding we have today is but a tiny fraction of what there is to be understood.

Various pundits have written about "the end" of such things as science, history, inflation, or whatever as if we had come to a watershed in our development. But myopia is not new. Nothing lasts forever, of course, but in a healthy society, as one avenue closes, others open up. Today, there is scarcely an aspect of life of which we can say, yes, we truly understand that. There is no doubt that we know more *facts* today than ever before. Assemblies of facts, however, as Henri Poincaré once famously pointed out, "are no more a science than a heap of stones is a house." Nowadays, countless committees evaluate proposals for the incremental contributions they might make to established fields, that is, to the status quo. Inevitably, proposals aimed at specific advances are favored at the expense of those that would concentrate exclusively on such intangible goals as understanding. Consequently, the scope for uncovering the new understanding that might generate new fields—such as molecular biology 25 years ago or "warm" superconductivity a decade ago—tends to be ignored. To mix some metaphors, it does not matter if you can see only the tip of an iceberg so long as you are not fooled into thinking that is all there is.

Long ago, Christopher Columbus had considerable difficulty in getting financial support for his first voyage of discovery—"the conquest of what appears impossible"—as he called it. Eventually, in 1492, the Queen of Spain agreed to sponsor him. Today, a modern Columbus would probably fail, not necessarily because his trip might be considered too risky, but because there would not yet be an "Americas" committee to consider his proposal.

Serious Problem 3. Under the present arrangements, the demand for research funds is much greater than supply. Far from being ignored, this problem is, of course, the magnificent obsession of scientists everywhere. Funding agencies seem to believe that if only they had enough money, their problems would be solved. Money alone, however, will not solve Serious Problems 1 or 2.

The obsession with Serious Problem 3 leads to some very curious logic. Most of us are accustomed to the democratic way of life. It might be natural, therefore, to assume that the democratic procedures can safely be applied to research selection. They cannot. Democracy is founded on equality and justice but tends to take the short-term view. Unfortunately, when committees strive to be fair to everyone, as they usually do, they tend to concentrate on "next-step" programs, that is, research which consensus agrees would make the most important contribution to the evolution of a field. However, these next-step, incremental options are rather like pebbles on a beach—no matter how many have been turned over there is always an effectively infinite number remaining. One cannot assert, a priori, that a particular step will not lead to something interesting or new. One cannot argue convincingly, therefore, that it should not be taken or that the funds required could be better spent. Concentration on next steps, therefore, creates the need for bottomless pockets to finance the search for them.

Next-step research also leads to competition and, in turn, to increased demands for ever more expensive equipment merely to stay in the game. Such leading journals as *Nature* and *Science* now abound with full-page advertisements in glossy color. One must presume that since they appear in those prestigious journals, they are addressed to serious-minded scientists, but it sometimes beggars belief as they exclaim such nonsense as this or that piece of equipment will "get your results faster" or "help you finish first" or "accelerate your discovery efforts." It is an extraordinary development.

On the national scale, the United Kingdom's research-and-development (R&D) expenditure is only some 5% of the global total. It has been plausibly concluded, therefore, that we cannot afford to do everything. Thus, it has been agreed that the United Kingdom should focus its R&D programs on areas that might make the biggest contribution to the national interest. The United States, however, accounts for roughly half of the global expenditure on research, but it is still not enough to satisfy the demand. The U.S. agencies too have concluded that we cannot afford to do everything, and scientists must also justify every dollar requested in terms of some desirable outcome. The per-capita research spend in the United States is roughly double that of the United Kingdom, but the shortfall has led to similar solutions. This would seem to indicate that money is not the main problem.

Up to the 1970s, universities in Britain were mainly autonomous.

Research was a matter of individual choice. Some academics did none at all, and there was little or no central planning. However, we should remember that even in those enlightened times scientists were not always free to do as they pleased. The chances are that a young person starting out would become a junior member of an autocratic professor's research group and more often than not would have little freedom in choosing what problem to work on. This was especially true at the turn of the twentieth century. But there were powerful professors everywhere, and not all of them were autocratic. They all had their unique perspective on what was important. The system was diverse and flexible enough for creativity to flourish. And it did, of course, as it always will given the chance. Nevertheless, one of the apparent drawbacks of this somewhat cavalier system was that some public money was wasted. There was little direct competition for funds, little or no control on their use, and sometimes some of them were frittered away.

On the industrial front, many corporations used to host wide ranges of research programs of which any university today would be proud. This is not the case today. Over the past decade or so, companies have focused* increasingly on core business, that is, on those sectors where a corporation has been most successful in the past. If that policy were universally adopted, it would of course be a sure recipe for economic stagnation in the long term as new interests could never be explored. Hopefully, enough up-and-coming small companies will buck that trend, and the danger can be averted. For the large technological companies, however, the effects have been more immediate. Exploratory research is necessarily unpredictable. It has therefore never been a substantial part of industrial R&D programs in general, although in chemicals and pharmaceuticals it was often more than 10%. Some companies such as IBM and Bell Telephone Laboratories (now Lucent Technologies) and BP, of course, the sponsor of Venture Research (see Chapter 8), were renowned for the breadth of their research interests and for their exploratory research in particular. See Text Box 22. However, the probability that the unpredictable benefits of exploratory research will fall within a company's reduced operational range and shorter time horizon is now so low that it has virtually disappeared from the industrial scene. To make matters worse, even the range of mission-oriented research has been cut back in line with companies' more focused operational requirements. Derivative products are all the rage because they can be produced without having to stray too far from old hunting grounds. They have the enormous advan-

*It has become fashionable to describe decades by some presumed characteristic—the swinging 60s, for example. I would suggest "focused" for the 1990s.

Text Box 22: Altruistic Sponsorship: IBM

In the 1960s, IBM's chairman, Thomas Watson, Sr., set up the Fellows Program. He appointed Fellows for five years to be "dreamers, heretics, mavericks, gadflies, and geniuses." He called them his "wild ducks" and gave them the freedom to call on support from mainstream activities. Their remit was simply to "shake up the system." Some of their successes must have been beyond even his wildest dreams. Working in the 1970s and early 1980s at the IBM Zürich Laboratories, two Fellows, Heinrich Rohrer and Gerd Bennig, developed the scanning tunneling microscope, a device that exploits the ability of particles like electrons to overcome barriers that classically would be impenetrable—as, say, a 100-foot wall would be to a high jumper. The process can be understood only by using quantum mechanics. The new type of microscope had unprecedented resolution and could provide images of subatomic structures. It transformed microscopy, and Rohrer and Bennig won the Nobel Prize in 1986. Another successful Fellow was Benoit B. Mandelbrot, who worked at the Thomas J. Watson Research Center in New York. He discovered "fractal" geometry, a word he coined, which quite literally introduced new dimensions to scientific study. Indeed, his work founded a new branch of mathematics and gave rise to a new type of art. His discoveries had, in principle, a breathtakingly simple basis—on his observation that the number of distinct scales and natural patterns exhibited by such objects as trees, clouds, or coastlines are in effect infinite. Remarkably, IBM's Zürich Laboratory triumphed again in 1986, when Georg Bednorz and Alexander Müller discovered "warm" superconductivity, although by then there were signs that some of IBM's more enlightened and altruistic policies were coming to an end as companies everywhere became more focused.

tage that they fit within core business strategies, and they lead to short-term competitive advantage.

As if all this were not enough, industrialists today have a strong influence on the selection of priorities for academic research. Therefore, academic scientists have to cope with the double whammy of government policies on focus and relevance together with the contracting scope and time horizons of industry.

In the good old days that ended about 15 years ago, research directors in large-scale industry everywhere were responsible for selecting the

Text Box 23: Altruistic Sponsorship: Bell Labs

Bell Telephone Laboratories were also dramatically successful. John Bardeen, Walter Brattain, and William Shockley discovered the transistor there shortly after World War II. Later, while using an ultrasensitive microwave receiving system to study signal reception from communication satellites, Arno Penzias and Robert Wilson found an unexpected background of radio noise with no obvious explanation. It came from all directions and, after repeated checks, it appeared to be coming from outside the Galaxy. They might have dismissed the faint "noise" as a slight nuisance of no practical importance, especially if they had been under the relentless pressure to deliver something exploitable as industrial scientists are nowadays. But they had the freedom to persevere for as long as it took and the prescience to realize that they might be on to something important. They eventually concluded that this whispering radio noise was perhaps the faint echo of the closing stages of the Big Bang. (This was the deliberately ridiculous name given by Fred Hoyle, a British cosmologist, to the creation of the universe some 14 billion years ago. Hoyle was frequently scathing about the very idea of a "moment" of creation.) Thus, Penzias and Wilson's experiment became one the most momentous of the twentieth century. All these Bell scientists won Nobel Prizes. Indeed, a total of 11 Nobel Prize winners came from the Bell Laboratories.

research that might lead to new and profitable business. They often had tremendous freedom of action, could indulge their flair for spotting a new trend, and could authorize new programs as they saw fit. Some were notoriously autocratic, but in common with the professors at the turn of the twentieth century, not all were, and in any event, they all had their own individual views on what was important. Nevertheless, one of the apparent drawbacks of this somewhat cavalier system was that some private money was wasted. Accounting practices were feeble compared with today's, so the small drain on resources usually went unnoticed. However, creativity flourished. See Text Box 23.

Indeed, all this apparent profligacy in both the public and private sectors led to a rich harvest of new technologies and enormous increases in economic growth. Research was clearly seen as a good thing, and the enterprise expanded. But it was not only its size that grew. Expansion was also accompanied by the increasing use of such words as efficiency, harmonization, rationalization, prioritization, and so on, in short by red tape. Red tape is hardly new. Bureaucrats have always loved it because it can be used to strangle those who step out of line. Determined people could usually find ways to avoid it. At the Battle of Copenhagen (1801) Nelson was told of a signal flying from his cautious flag officer's ship ordering him to withdraw. Nelson apocryphally put the telescope to his blind eye to give him a witty excuse for his lapse. He ignored the signal and went on to win.

Ironically, however, scientists have handed the bureaucrats a priceless weapon. The transistor and an avalanche of subsequent developments led to computers of growing, easy-to-use power and declining cost. Consequently, a single person can oversee the entire operations of an organization of virtually any size and in virtually any detail. Spreadsheets, for example, can compare the efficiency with which every employee executes every duty or with which every resource down to desks or laptops is used. Even the Nelsons might not now evade the bureaucrats' grip. To abuse a well-known Churchillian quotation:

> Never in the history of intellectual endeavor have so many ideas been strangled by the efforts of so few.

It is no longer acceptable today to have *viable* procedures for doing something. One must now be able to show that the resources required will be used in the most efficient way at all times. It has been forgotten that a little inefficiency in human affairs can be a good thing. A good employer will recognize that a laborer leaning on his spade is not necessarily shirking but may be drawing breath before the next onslaught. Top-flight long-distance runners vary their pace to improve their overall performance. Scientists too must be allowed to please themselves at times. The secret lies in knowing which ones should be given that sort of freedom.

Until about 1970, scientists were virtually free to use what little funds they could get hold of as they thought fit. With the benefit of hindsight, we can now see that the intellectually freewheeling research of the preceding decades resulted in a wealth of new sciences and technologies. Economic growth soared. The new industries created included the following:

- Semiconductors and integrated circuitry
- Lasers and optoelectronics
- Jet engines
- Nuclear power
- Computers

- Photocopiers

- Polymers and plastics

- Robots and automated production

- Genetic manipulation and biotechnology generally

These developments were not refinements of earlier industry. They were not predicted. Few of these discoveries were seen as being needed before work on them began. We might wonder, therefore, how many of the pioneers* who made them would have satisfied the demands of such initiatives as today's Foresight in the United Kingdom had they been fashionable at the time? Luckily for almost everyone in the advanced countries, they did not have to.

Economists generally describe the period between 1950 and 1973 as the "Golden Age" because of its high growth rates. Angus Maddison (1989) discusses the possible causes in *The World Economy in the Twentieth Century*, from which excellent source we drew in Chapter 4. He concludes (p. 34):

> The acceleration did contain some elements of recovery but it lasted so long and involved such a massive and sustained improvement in performance that new forces were clearly at work in the growth process.

Maddison knows, of course, that technology is "the major engine of growth," as he describes it, and so his remark is particularly interesting. Furthermore, he does not say what the new forces might be.

My conjecture is that Maddison's "new forces" stemmed from the outstanding *quality* of R&D in the decades preceding the Golden Age. The industrial R&D laboratory emerged around the beginning of the twentieth century. See Text Box 24. Research until then had been largely an academic province, and so the *concept* of industrial R&D and its relationship with academic research had also to be developed. In particular, it

---

*Charles Townes won the Nobel Prize for Physics in 1964 for his contribution to the discovery of the laser. In 1999, he published *How the Laser Happened: Adventures of a Scientist* (Oxford University Press, New York). He stressed that the essential steps in the development of the laser derived from scientists "playing" and freely exchanging ideas and from fortuitous opportunities realized and put into use by open-minded researchers. No strategic planning by expert panels could have led to the development of the laser as it actually happened. Indeed, many experts (including such Olympians as Neils Bohr and I. I. Rabi) were convinced that the laser could not work.

> ## Text Box 24: The First Industrial Research Laboratories
>
> Although the Industrial Revolution began in Britain, technological leadership had moved to the United States by the end of the nineteenth century. Innovation did not dry up in Europe, but the United States became better organized to adopt and develop new ideas from whatever source they came. In 1900, General Electric, for example (which can trace its origins to Edison's Electric Light Company), adopted what was thought to be a revolutionary idea by, according to GE's own account, being the first company to establish its own industrial R&D laboratory. GE's laboratory was possibly the most influential in providing a model for others, but Siemens, the German company, had an industrial research laboratory in 1869. Others may also have been in place by the turn of the century—see Moses Abramovitz (1991).

had to be learned that although a manager or shareholder may propose, it is for Nature to dispose. Not every problem may be economically soluble.* By the 1930s, however, these laboratories were becoming very efficient. The following decade included a world war, the outcome of which was strongly influenced by technology. War is a terrible thing, but it can awaken a "can-do" philosophy. It can kindle a spirit of determination, not only among scientists, but also among those who provide the resources and create the environments in which scientists work. The eventual result was the rather staggering list of *new* technologies listed earlier.

Unfortunately, shortly after the research enterprise had delivered its most recent earth-shaking crop, its controllers began to develop the policies they thought would put the enterprise on a more rigorous footing. Perhaps the most significant idea to emerge was that the geese that laid the golden eggs were too valuable to be allowed to roam freely. Thus, in stark contrast to the relatively tolerant environments of the run-up to the Golden Age, we began to forget that pioneers must have freedom. If we want a healthy society, this condition should always be satisfied, particu-

---

*Recall Francis Bacon's profound remark, "Nature is only to be commanded by obeying her" (p. 52), a subject to which I have promised to return in the next chapter.

larly when the going gets tough. If it is taken as an option that can be postponed to some future, less straitened times, those times may never arrive. But these obvious lessons from history are being forgotten, and the research enterprise is being managed and controlled more strictly than ever before. The policies that imposed the need for creativity to be *cultivated* were sold on the assumption that it would be more productive than the untamed, wilder variety. More than 20 years of tight management later we ought to be able to see signs of an improved crop, especially as resource allocations have been rising.* But where are the new sciences that were not created in the relatively lean and freewheeling years? What unpredicted discoveries have been made that could eventually lead to substantial economic growth?

It may be too early at present to look for full justification of the new policies. For most scientists, however, two or three decades are a very large slice of a productive life. We should therefore be able to see signs that the rigorous imposition of quasi-farming methods on the management of creativity are yielding more successes than the relaxed regimes that tolerated hunter-gatherers. My tentative audit of the harvest so far from this well-managed generation is as follows:

- High-temperature superconductivity

- Carbon-60, fullerenes, etc.

- ?

The discoveries on this shortest possible list, however, would seem to have come about *despite* the new arrangements. See Text Box 25. In 1986 the Swiss scientists Georg Bednorz and Alexander Müller discovered a new class of high-temperature superconductors by turning their attention to a class of materials more noted for their *insulating* properties—the ceramics, rather than the metals on which all previous attempts had concentrated. They were working at the time at the IBM Zürich Research Laboratories, but it has been widely reported that they were acting on their own initiative. Indeed, it seems to have been necessary to conceal what they were doing "behind the fume cupboard," to use the euphemism for programs outside the officially approved activities. They won the Nobel Prize for Physics in 1987. Carbon-60 was discovered by a group of

---

*In the United States, for example, the total federal and industrial investment in R&D over this period has more than doubled in real terms with smaller increases elsewhere (although they still fall far short of the demand).

Text Box 25: A Japanese Story

In 1989, the Anglo-Japanese High Technology Forum meeting in Gotemba, Japan, discussed creativity. Someone asked why almost every major scientific discovery had come from the West while Japan had been more successful in developing Western ideas. A Japanese delegate perhaps jokingly remarked to me that these facts should not be surprising as the tradition of hunting, which favors the aggressive individual, is strong in Britain and other Western countries and could explain their success in original research. Japan, on the other hand, was really a nation of farmers with a long tradition of working in harmony with the land and with each other, hence their success in development. So why, he asked, were the British trying to destroy their traditional advantage by giving so much weight to consensus?

---

equally determined scientists, one of whom was partly self-supported.* They also won the Nobel Prize.

Nevertheless, our current perspectives on today's world cannot include hindsight, and so my short list may be extended later. Biotechnology, for example, is becoming increasingly important. It has had many technologically productive years, and such magnificent techniques as gene sequencing and the polymerase chain reaction have transformed prospects. But there has been a dearth of scientific breakthrough. No "Newton for the stalk of grass" has yet emerged, to use Immanuel Kant's delightful phrase. My questions may be somewhat premature, but should we be satisfied with what has been achieved so far by policies that dragoon scientists into being more purposeful?

All Organisation for Economic Co-operation and Development (OECD) member states have sophisticated scientific and technological infrastructures. However, in a leader of July 2, 1994, *The Economist* said:

---

*The British member of the trio to discover ^{60}C—Harry Kroto of the University of Sussex—personally had to support some of the crucial early stages of the work. He shared the Nobel Prize for Chemistry in 1996 with Robert Curl and Richard Smalley of Rice University in Houston. In the six years before the Prize was awarded, Kroto had a string of applications to EPSRC for substantial funds turned down—one on the grounds that the research was too open ended!

The oddity of modern times is that it has become so much harder to make a product that is genuinely different or better than a competitor's. And if you do make one, it is harder than ever to stay ahead. One reason is that technical expertise is more widely held by a bigger universe of companies, and is disseminated at ever faster speeds. Make a better mousetrap and the probability is that, within a few weeks, a cheaper version will appear next to yours, having been reverse-engineered in Hong Kong and assembled in Guandong province.

Six years later, in another leader, *The Economist* (October 7, 2000) quoted the then U.S. Treasury Secretary Larry Summers:

[I]nnovation is increasingly driven by firms that win temporary monopoly power but enjoy it only for a moment before being replaced by a company with a better product that itself gains a short-lived monopoly.

Generic technologies, on the other hand, are not so vulnerable. If we go back to the early 1960s, say, the reverse engineering of a transistor by would-be competitors in low-wage countries would have been impossible because electronic expertise would have been in thermionic valves. Today, however, there have been no genuinely new technologies (as opposed to derivatives) for some time, and the very few scientific breakthroughs of recent times have yet to realize their potential. Industrial R&D policies, therefore, are now in effect based on intellectual asset stripping. They are based on seminal discoveries made decades ago, and little in the way of new generic intellectual capital is being created.

These conclusions may seem bizarre. Technological progress seems faster than ever, particularly in electronics and communications—the so-called New Economy. However, the durability of its prolific products was questionable even before the dotcom mania suffered its inevitable demise. One should not go out and buy a personal computer, say, because a more powerful version will be available at the same price or less in a few months, a prevarication that could go on indefinitely. Computers are bought, of course, but their resale value plunges almost immediately. The Internet is both a major opportunity and a special source of concern. Communication has never been easier, but how should its benefits be valued? How do they contribute to economic growth?

To illustrate the problem: An Internet company was set up in 1999 to trade specialized steel products. Within a few months, it was worth as much (in January 2000) as the entire U.S. steel industry. There are similar stories in abundance. Now, if a steel company valued at, say, $100 million bought another steel company of the same valuation, their combined

---

Text Box 26: A Japanese Joke

Akio Marita (1921–1999), the cofounder of the Sony Corporation, was frequently scornful of the ways the leaders of the industrialized nations were ignoring the lessons of history. He told me the following tongue-in-cheek joke in the late 1980s. The captain and senior officers of a large ship are playing poker in the captain's cabin. A crewmember knocks on his door to say that the ship has been holed below the water line and is sinking. "Go away," says the captain. "Don't you know we're playing for high stakes?"

---

value would be some $200 million. Such a blindingly obvious statement may seem unnecessary, but remember we are about to reenter the Alice in Wonderland world of the New Economy. If the Internet company in the illustration had *actually* used funds raised against its valuation in January 2000 to buy the U.S. steel industry, what would their combined value have been? Immediately after the hypothetical takeover, a buyer would see an integrated steel company selling products whose total value was unchanged. The new company would probably be more efficient at selling. Would the increase in efficiency have doubled the value of the steel? I doubt it. The illustrative company was worth real dollars. So, what is the relationship between the values of Internet and traditional companies?

The Internet is different from other major advances. The steam engine, the laser, integrated circuitry, nuclear power, and other generic discoveries resulted in new things being done that previously were not merely difficult; *they were inconceivable*. The Internet allows presently conceivable things to be done with unprecedented efficiency. As today's world is obsessed with efficiency, the Internet must seem like a gift from the gods. And it is, if we understand what it is. See Text Box 26. One of the Internet's characteristics is that it permits new combinations of existing things, such as playing a computer game and following the Stock Exchange at the same time, thereby creating the illusion that value has been added. The illusion is so powerful that it can be traded for real money, at least for a time, thereby inflating the economic growth statistics.

The Internet is an efficient means of exchanging and transporting information. Consumers can easily locate the cheapest products, therefore, and manufacturers must endure a steady erosion of profit margins. In favoring consumers over producers in this way, the Internet is not typical of other generic advances. The "oddity of modern times" quoted above arises because advanced-world companies are not taking advantage of their potential strengths. The Internet, among other things, accelerates the diffusion of mature technologies (and their derivatives) and puts advanced-world companies at a severe disadvantage with respect to the developing world because of its lower labor and infrastructure costs. Should it be surprising, therefore, that global per-capita economic growth is declining? Are the controls now being applied to R&D inhibiting the creation of genuinely new technologies? Are these controls deactivating Maddison's "new forces" and reducing the prospects for growth from those enjoyed in the good old days of *laissez faire*?

The faster industry uses up intellectual capital, the more it will appear that we are experiencing healthy growth. This would seem to account for the rosy economic outlook of some advanced countries at the turn of the century. It could not last, of course. The New Economy, with its efficient and multitudinous avenues of communication, accelerates the consumption of intellectual capital in the fields from which it draws. If the portfolio is not expanded, preferably with generic technology, the New Economy brings diminishing returns and puts us on a fast track to economic stagnation. That would be very unpleasant if population continues to rise at its present rate. Scientists created the Internet to help them handle scientific information. Used imaginatively, it is a powerful resource. Scientific breakthroughs not only rejuvenate economies but also restore the advantage to the producers who exploit them. The harvests of a healthy scientific enterprise can thereby defer stagnation indefinitely.

Professional economists have yet to resolve these and other issues raised by the New Economy, but I shall return to some other aspects of the nature of quality in the next chapter.

Technology enables companies to make things, but economic activity is governed by consumption and trade. World trade since 1950 has increased at a very much faster rate than GDPs have increased—see *The Economist*, October 3, 1998, p. 5. Trade also diffuses technological know-how as companies copy and imitate. For a given level of intellectual capital, therefore, trade should become self-limiting. The more nations trade, the more they reduce comparative advantage, and the less they will want to trade in the future. Breakthrough technologies automatically generate advantage, of course, because they create new intellectual capital. In their absence, however, as the old technologies mature and diffusion becomes more widespread, one might expect a progressive harmonization of capabilities and per-capita production rates. Figures 11 and 12 seem to indicate that real GDPs per capita for the industrialized countries have indeed been strongly converging since about 1950. Is there indeed a pow-

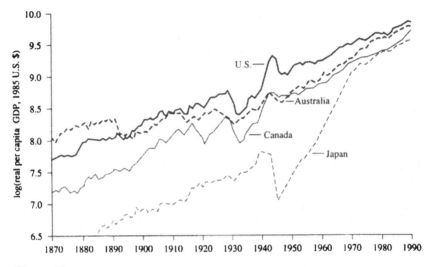

Figure 11
Real GDP per capita in 1985 U.S. dollars for the United States, Canada, Australia, and Japan. (Reproduced by permission from R. J. Barrow and X. Sala-i-Martin, *Economic Growth*, McGraw-Hill, New York, 1995.)

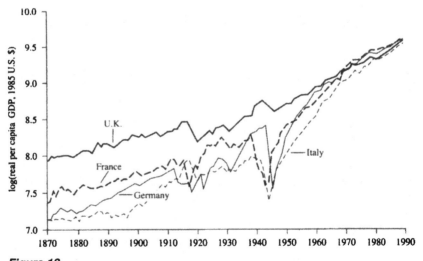

Figure 12
Real GDP per capita in 1985 U.S. dollars for France, Germany, Italy, and the United Kingdom. (Reproduced by permission from R. J. Barrow and X. Sala-i-Martin, *Economic Growth*, McGraw-Hill, New York, 1995.)

erful feedback mechanism that progressively reduces the inclination to trade as capabilities are harmonized?

Effects and their causes are often difficult to connect in economics, but the sustained decline in the average per-capita growth rate requires an explanation, not to mention a solution. It seems inconceivable that the bureaucratic trends outlined in Chapter 5 will not have had an adverse effect on the science-and-technology supply and, hence, on growth. Science has not come to the end of the road; its potential is as vast as ever. The problems are mainly structural, and their most important element is the virtually all-pervasive peer-review bureaucracy—that highly dispersed latter-day inquisition for the defense of orthodoxy to which every scientist must now submit. The rack may no longer be threatened, but suspension of funding and excommunication from a scientific community are still possible for those who would seriously challenge the status quo. See Text Box 27.

Peer review is not necessarily bad—the problem lies in its all-pervasiveness. If there were more flexibility, more pioneers would be encouraged to push their ideas forward. Economic vigor would also get a boost because an increased number of radical breakthroughs would lead to more surges in economic growth. Unfortunately, the rare blooms of breakthrough are, like orchids, found only in the wilderness. That has al-

---

Text Box 27: The Social Pressures to Conform

Those who think the Inquisition analogy is fanciful should read the book by Jessica Wang (1999), a history of American science in the period of the 1950s leading to the trial by the Atomic Energy Commission of Robert Oppenheimer in 1953 to 1954. Oppenheimer led the team that built the atomic bomb in 1945, a scientific and technological *tour de force*. Nevertheless, he was stripped of his security clearance, a verdict that ended his career. Wang relates how the overwhelming majority of scientists chose not to question the system of anticommunism, often referred to as McCarthyism, with its accusations of "un-American" activities, and indeed tacitly accepted the need for that Inquisition. Furthermore, official policy in the United States during the McCarthy period was to terminate the research grants of scientists *suspected* of disloyalty to the U.S. government. To its great credit, the then newly established National Science Foundation did not follow official policy and went as far as to provide grants to those who had been cut off by other agencies. Linus Pauling was one such beneficiary.

ways been their habitat. But in today's risk-averse world, would-be pioneers must be extraordinarily persistent or very imaginative in their bids for funds to allow them to leave the well-beaten tracks. That is no way to reach the unknown.

A recent discovery in archaeology illustrates the situation very well. When rain forest is cleared using normal methods, much of the organic matter disappears under the action of sun and rain, leaving a poor acid soil low in nutrients. Work in the late 1950s exploring the Amazon Basin for signs of old civilizations concluded that the soils were so poor that they could not provide sustenance for significant numbers of people. As this is the state of most of the soil in the Amazon Basin, this conclusion brought the searches for materially advanced cultures to an abrupt halt. It would have been pointless to fund exploration in areas known to be barren. However, an international group of scientists were not convinced and recently formed the Central Amazon Project (see *Science*, August 9, 2002, p. 921). Their work has revealed the existence of extensive settlements of ancient peoples who lived up to about 1000 years ago. Remarkably, these peoples found ways of developing a soil, now known locally as *Terra Preta do Indio* (Indian dark earth) that is rich and very productive. Apparently, these soils were used for centuries without losing their high levels of productivity.

These peoples disappeared long before the Spanish came, but astonishingly, their fertile soils have survived the centuries. They are still rich in organic matter and mineral nutrients such as nitrogen, calcium, potassium, and phosphorus and are ready for immediate use. They are typically found in plots of a few acres, but some cover 600 acres or more. Now that scientists know what to look for, the soils are being found in abundance.

The Indians' secret, apparently, was to mix indigenous charcoal into the soil, but microbial organisms also seem to be important. Research on *Terra Preta* aimed at discovering how the ancient Amazonian Indians worked their wonderful trick has only just begun, but the soils made so far are giving very high yields. The technique should be generally applicable to hot regions, particularly those being deforested. The positive implications of this discovery could therefore be enormous, especially for the developing world. However, it could have been made decades ago had it not been for the veto of consensus.

This example is in archaeology, but similar stories could be told in every discipline, as we discovered with our Venture Research project (see Chapter 8). We should give *total freedom* to some pioneers, but how can we reliably identify those who should have it? There are many scientists and many fields. One does not even know which haystack hides the needle. It has been forgotten that we did not have to have special arrangements for finding the Einsteins in the past. There was enough flexibility in the system to allow them to emerge, but that has been taken away in the quest for efficiency. Nowadays, the consensus seems to be that it is impossible to spot the Einsteins in advance. Search procedures would therefore be a waste of time, and so *no one has ever looked for them.*

A new process for identifying pioneers—self-selection—is described in Chapter 8. Once the need for flexibility has been recognized, other efficient processes will emerge. The problem is becoming urgent. Industrialists and investors might wish to reflect on the long-term consequences of a strict adherence to core business strategies combined with the predictable products of the academic peer-review bureaucracies. Then perhaps the most adventurous among them will take the appropriate actions.

Science and technology are not the only enterprises losing their diversity. Established in 1995, the World Trade Organization (WTO) had 146 member states in 2003. Its rules apply to over 90% of international trade. National economies are becoming ever more closely linked. Regional trade agreements are also spreading. The WTO lists 76 free-trade areas or customs unions set up since 1948, and of these, more than half have come in the 1990s. The list includes the European Union (EU), whose member states seem determined to iron out any differences in the law, finance, safety, or any customs and practices* that it believes might impinge on competition among its membership. Similar unions have been set up in North America (The North American Free-Trade Agreement), South America (Mercosur), and the Far East (The Asia Pacific Economic Co-operation Forum).

Planned mergers between multinational corporations were announced almost daily during the 1990s. According to Securities Data Company, mergers were worth $2400 billion worldwide in 1998, a 50% increase on the 1997 figure, which was also a record year—see *The Economist*, January 9, 1999. On November 28, 1998, *The Economist*, in a leader discussing the reasons for the high merger rate, concluded:

> All too often it is a case of imitation; somebody else has done it, and so we should too. Or it is a generalized, defensive fear: globalization, say, or the arrival of Europe's single currency ... creating a strong incentive for the insecure to leap into bed with one another, the better to prevent their beds from being taken away altogether.

---

*The EU's infatuation with petty regulations is legendary. In 2000, for example, it became *illegal* in the United Kingdom to offer goods at the point of sale weighed in pounds or ounces. The kilogram is the only measure allowed.

One might add management consultants and business schools that create an enormous bloc of uniform thinking.* When I was employed by British Petroleum (BP) in the late 1980s, the focus on downsizing, delayering, efficiency, accountability, networking, and other fashionable buzz words suddenly became all the rage. My job (see Chapter 8) was probably unique and took me far and wide outside BP in both the academic and industrial worlds. It was most disconcerting to hear the many BP staff I respected using these fashionable words as if they had just invented them. This of course is a tribute to the consultant's art. My colleagues' main preoccupation seemed to be how they could best implement the results of their new thinking and thereby steal a march on their competitors. Unfortunately, all too often, their competitors were doing exactly the same things. The escalating merger rate was no doubt driven also by the very high stock market valuations of the late 1990s and has now (in 2003) reduced considerably. Nevertheless, according to Dealogic, a company specializing in mergers and acquisitions, there were more than 3000 mergers in the United States in the first half of 2002, worth in total some $186 billion.

In a most depressing vision of the global future, therefore, we will all reside in one of a few regional blocs, select our supplies and services from those which a few giant conglomerates agree to provide, and struggle to express our individuality in a mire of increasing regulations and controls. This gloomy, Orwellian prospect is not some fantastical science fiction scenario. It could come about in only a handful of decades if we continue more or less unchecked on our present path and, particularly, if we persist in indulging our current obsession with efficiency and short-sightedness.

There is an additional cause for concern. Several corporations have more economic clout than most of the world's 200 or so nation states. Indeed, of the world's 100 largest economic entities, only half are nation states, the rest being corporations. On January 9, 1999, *The Economist* light-heartedly reported a conversation on the planning of a particularly large merger (p. 21). "Can anybody stop us?" someone asked, to which the reply had been "NATO." There seemed to be no sinister implication in the story, but if a supercorporation were to embark upon the extension of its influence by other than commercial means, it would not be the first. The enormously successful English East India Company, for example,

---

*In 1998 one of the largest management consultants employed some 40,000 globally, all so intensively trained in the company's methodologies that their competitors referred to them as androids. In 1994, there were six management consultant companies with revenues greater than $1 billion a year.

founded in 1600, had its own army and navy and frequently used them both in its long history without much scruple.

The gloom is not inevitable. Scientific breakthroughs not only transform economic prospects but can also offer some protection against the unchecked expansion of the power and influence of supercorporations. *Science can divide and rule.* This perhaps surprising property stems from the fact that large corporations today find it almost impossible to grow new initiatives that might take them into new types of markets. They have the people who could do so, but if they are to be successful, they would need freedom and flexible sources of finance, a stake in their enterprise, and rich rewards if they did succeed. Large organizations today seem to find it impossible to advance such latitude to the would-be pioneers among their staff, especially if they are relatively junior. Highly efficient accounting procedures and focus on core business also mean that there is nowhere to hide expenditure on extracurricular activities "behind the fume cupboard" that might lead to tomorrow's industry. But if investors were to restore their risk tolerance to its former levels, successful pioneers could take their breakthroughs forward in small companies, some of which, like little acorns, may eventually emerge as larger entities to balance the hegemony of the supers—and so on. That's how healthy growth should proceed.

In an essay in *The Economist*, the celebrated management consultant Peter Drucker (2001) discusses the future of work and of the corporation. In his concluding remarks, he says (p. 21):

> The next society has not quite arrived yet, but ... to survive and succeed, every organization will have to turn itself into a change agent ... This requires the organized abandonment of things that have been shown to be unsuccessful, and the organized and continuous improvement of every product, service and process within the enterprise. It requires the exploitation of successes, especially unexpected and unplanned-for ones, and it requires systematic innovation. The point of becoming a change agent is that it changes the mindset of the entire organization. Instead of seeing change as a threat, its people will come to consider it as an opportunity.

I applaud his remarks. However, his essay does not discuss current corporate obsessions with short-term efficiency and rigorous accountability and the almost universal policy among larger companies of focusing on core business. These newly acquired mindsets seem nevertheless deep rooted, perhaps because they resonate with our natural dislike of change. But it is imperative that the necessary changes are made soon. They would not need a revolution, but sometimes even much needed minor change can be resisted passionately. There are many examples, but to take only one: Even after the discovery in 1753 by the British naval surgeon James Lind that small amounts of lemon juice cure scurvy, which was then a major cause of death at sea, it was some 40 years before the British Navy could be persuaded to authorize the necessary change to seamen's diet. We now know that a small intake of vitamin C, which of course is found in lemon juice, is essential to life. After Robert Solow's discovery, we must also realize that providing just a little freedom to a precious few pioneers will have a similar effect to vitamin C in a healthy diet. Without regular shots of that vital ingredient, the patient—civilized life in this case—will soon fade away.

7

Re-Creating
the Golden Age

My hypothesis is that the recent changes to funding policies are stifling creativity and reducing global economic growth. It may be proved wrong, but fundamental problems would remain. There is no question that science and technology are the dominant sources of economic growth. In recent years, however, we have changed the ways in which new science and technology are acquired without considering the effects these changes might have on this invaluable economic relationship. My suggestion is that the changes have diminished the potential of scientific enterprise. Creativity cannot be caged. I suspect, however, that economists might dismiss my offerings with some justification as arm waving, insufficiently rigorous, or incomplete. My task in this chapter will be to try to introduce a little more precision into my reasoning. I will try to explain how the surge in postwar growth came about and how it might be restored, and, hopefully, to persuade professional economists that their treatment of growth theory needs to be modified in the light of recent developments.

There is a voluminous literature on economic growth. One of the most powerful sources of growth—technical change—is still poorly

Pioneering Research: A Risk Worth Taking, By Donald W. Braben
ISBN 0-471-48852-6   © 2004 John Wiley & Sons, Inc.

understood, however. As Maddison points out, its measurement is still "most elusive." It should not be too surprising. Until only a few decades ago, economists were struggling to account for the enormous difference between the growth in labor productivity, and the growth of what had been thought to be the primary input—namely the capital per worker. In 1957, Robert Solow attributed the difference to an unknown factor he called "technical change" (p. 59). More recently, Edward Denison in his heroic study of the American economy* tried to quantify a wide range of factors that might contribute to growth, including education. He concluded that seven-tenths of the increase in output per worker was still left unexplained. As his Final Residual proved to be nearly constant during the period 1948 to 1973, he regarded it as a measure of growth due to the "advance of knowledge" incorporated into production.

I suggested in the last chapter that the mystery surrounding the decline in growth would be dispelled if account were taken of the *quality* of investments in R&D as well as their magnitude. Quality should be expected to have a strong influence on the nature of the output, namely the advance of knowledge, and consequently on economic growth too.

My use of "quality" in this context needs to be amplified. Thus, the quality of an investment in science and technology is related to the probability of achieving the investor's goals. High-quality investments require an intimate understanding of the current state of scientific or technical knowledge and of the ways in which the limitations imposed by that knowledge may influence the required outcome. Investors may also affect quality by overly specifying objectives or introducing extraneous factors. In this context, quality is not related to the potential value of the goals to be achieved or of the ability of the scientists involved. For example, a supremely rich investor who for personal reasons commissions a large team of the world's best scientists to find a cure for, say, cancer or AIDS within 10 years would be making a poor-quality investment. On the present levels of understanding, the likelihood of success would be remote even if the resources provided were unlimited. More prosaically, a car manufacturer who instructed its R&D laboratory to deliver a specified and radically improved mileage performance—say, a 100 miles per gallon, low emissions, high acceleration and maximum speed, and tight controls on retail price—all to be delivered in, say, two years would also be making a poor-quality investment. In this latter case, such detailed instructions would allow the technical staff almost no flexibility, freedom, or time to exercise their ingenuity. It would be virtually certain to fail. On the other hand, the car manufacturer could make a high-quality invest-

*Edward F. Denison, (1985).

ment by commissioning an improvement—any improvement—in performance or cost of production of a current model or a new one that would lead to an increase in sales and profits. Full rein would be given to creativity, and technical staff would probably come up with something that would make money.

The quality of investments is affected by the likely durability of the chosen objectives, as implied in the last chapter. Thus, investors trying to grab a niche in an existing technological market may soon find that their competitors have either matched them or can easily catch up or imitate them.

Attitudes to risk can also affect the quality of an investment. In the everyday world, if one wishes to reduce exposure to risk, one buys insurance—against fire, for example. It is a depressing fact of life, however, that one's enthusiasm for doing everything to prevent fires will be somewhat dampened if one will be fully compensated. Indeed, any insurance policy or device for reducing exposure to risk has the effect of reducing the incentive (Arrow 1962) to cope with it oneself. Research is a form of insurance against the consequences of an unpredictable future, but all research involves risk. Policy makers are faced therefore with the difficult question of how that risk should be dealt with. A possible option is to accept the risk, whatever it may be, and manage it in such a way as to promote a positive outcome. One should be prepared to respond to whatever difficulties or opportunities might emerge, but even if one has extensive knowledge of the research enterprise, the uncertainties cannot be removed. Indeed, researchers should be given every possible incentive to take calculated risks. *They* must make the calculations, of course, and so research under a "managed-risk" regime is intensely individual.* They should also choose the objectives, and so it is highly likely that there will be no competitors, at least initially. The premiums for this insurance are relatively low. If only highly motivated researchers are chosen, the research offers a high-quality investment as it is likely to create new intellectual capital. The outcomes of each specific investment are unpredictable, of course, but if this option is widely adopted, it offers the most comprehensive insurance against an unpredictable future that human ingenuity can provide.

Another option is to try to eliminate risk and hope thereby to have little or no risk to manage. This option has become the accepted norm almost everywhere. The ubiquitous idea is that risk can be minimized by focusing research on attractive products for which there is a commercial, social, or intellectual demand. Thus, one tries to arrange for a surprise-

---

*One such regime is described in more detail in Chapter 8.

free future, which of course would be the best insurance one could get—*if it worked!* The most attractive products are those that can be delivered soonest, and so the option favors short-term research. The option is also the most expensive, as the competition for attractive products not only is fierce but also increases as intellectual capital is used up. Expense is not the only disadvantage. Every scientific breakthrough we now enjoy came from the unknown. Uncertainty is an essential characteristic of this domain. But this supposedly minimal-risk option offers no incentives to scientists who believe they can manage the risks of straying far from the beaten tracks. Indeed, they are unlikely to get funded. This option is therefore likely to lead to poor-quality investments and diminishing returns.

If competitive advantage cannot be gained through technology, perhaps because of the poor quality of company investments, other alternatives may become more attractive. Mergers have increased enormously in the last decade. Mergers, however, are at best a zero-sum game. They do not generate new products or new markets, and the scale advantages can easily be swallowed up by increased problems of coordination. Loss of morale can be an important offsetting factor too as staff realize that their jobs might be part of any economies of scale.

The quality of government investments in science and technology began to decline in the 1970s—the quality of industrial investments held up for another decade or so. As quality deteriorated, the range of Denison's "advance-of-knowledge" factor would also fall. More and more was being discovered about less and less as investments were increasingly focused on selected targets. It should not be surprising, therefore, that per-capita economic growth fell as a result.

The reasoning outlined above was developed over about 20 years. It was not a theoretical exercise. As I will describe in the final chapter, it was used in a modest initiative that set out, in effect, to make high-quality investments in research, although that terminology was not used at the time. It was industrially sponsored, operated in Europe and the United States, and ran for 10 years. It was very successful. The reasoning should therefore have global implications and contain at least the germ of a possible solution to our economic problems. However, it should also be able to explain why economic growth rose to the high levels of the postwar years. We *know* what happened at that time, of course, and it is difficult to avoid the mental traps so nicely expressed in the Latin phrase *post hoc, ergo propter hoc* ("after this, therefore because of this") and attribute the exceptional growth to our favorite combination of events.

Growth theory, as *The Economist* has so succinctly described, is "the blackest of black boxes," and there is no generally accepted explanation for the postwar surge. I would offer the following.

Bacon's dream of systematically harvesting science for potentially valuable crops was finally realized in the twentieth century. There were very few industrial research laboratories at the end of the nineteenth century, but their number burgeoned at its turn. The United States led the way and, in doing so, took the vital first steps to becoming the new superpower. In 1921, the U.S. National Research Council listed 307, but by 1960, that number had risen to 5400. Almost a half of these "laboratories" oversaw the efforts of single individuals, essentially private inventors struggling to make their fortunes. Like Chester Carlson, the inventor of the photocopier (see Chapter 6), most would be hoping that a major corporation would foster the development of their brainchild. The cultivation of science as a business had to be learned, as had its relationship with technology. These processes are not always perfectly understood today, and it would be expected, therefore, that progress in the early decades would be slow. Moreover, the first half of this century was tumultuous—a world war was soon followed by a great depression and in turn by another world war.

The first half of the twentieth century saw some interesting times indeed. We had to struggle to learn how to exploit the apparently limitless potential of technology and to build the financial and social infrastructures that would regulate financial markets. Its second half reaped the benefits. There were no sustained and deep financial collapses and no world wars. Unfortunately, we are still beset by deep-rooted national, ethnic, or religious problems and their associated violence.

It is necessary to take a step backward in time to appreciate the reasons behind the Golden Age's surge in growth. Adam Smith was the father of economics and the modern treatment of economic growth. His *Wealth of Nations* (1776) inspired generations to follow. To quote the American economist Moses Abramovitz (1991, p. 5):

> Smith's theories were developed and refined in the decades after the appearance of his great book. Malthus's famous essay on population, taken together with Ricardo's treatment of diminishing returns in the use of land, sharpened the sense of conflict between population and resources. At the same time, there was a growing appreciation of the possibilities of progress based on the advance of knowledge. John Stuart Mill's *Principles of Political Economy* (1848) gave the economics of growth its definitive statement at the hands of classical economists.

Mill was one of the most enlightened and visionary writers of the nineteenth century. He defended laissez-faire economics but only if the rights of workers were respected. He argued that human rights should be restricted only where they impinge on the rights of others. He was among the first to advocate the feminist cause, maintaining that there was nothing in the many differences between men and women to justify the allocation of different rights. His views on growth, however, were especially far-sighted. In his *Principles of Political Economy*, Mill (1848, p. 696) said:

> Of the features which characterize this progressive economical movement of civilized nations, that which excites attention, through its intimate connection with the phenomena of Production, is the perpetual, and so far as human foresight can extend, *the unlimited growth of mankind's power over Nature.* [My italics.]

As Abramovitz (1991) has recently pointed out, it has taken later economists some time to regain Mill's sweeping view, if indeed they have.

With the notable exception of Joseph Schumpeter (1883–1950), the study of economic growth was neglected for 100 years after Mill's definitive work. Throughout this time, therefore, *economic* performance was rarely measured. Scientists and inventors were still judged solely on the quality of their eventual output. They could make hay while the sun shone and do as they pleased within the limits of their finances. In 1876 Thomas Edison (1847–1931) set up an "invention factory" in Menlo Park, New Jersey, employing some 50 people. With the supreme optimism of youth, he publicly declared that his laboratory would produce a minor invention every 10 days and a "big trick" every 6 months. The offering of such an outrageous hostage to fortune might have paralyzed a lesser person. In effect, "the wizard of Menlo Park," as he was soon to be called, was gambling on his ability to manipulate his understanding of Nature to produce practicable inventions to order. A workaholic, he applied for up to some 400 patents a year—1093 were granted in his lifetime. They included the phonograph, the electric lamp, the mimeograph, and the alkaline storage battery. He did not invent the motion picture projector, but his genius made it operational, and perhaps more than anyone he was responsible for creating the mass entertainment industry. He moved from Menlo Park in 1887 to build a facility 10 times larger in West Orange, New Jersey. The success of his laboratories* showed the world how human inventiveness could be made to be hugely profitable, and by the turn of the century they were being used as models for the industrial laboratories that would transform twentieth-century life.

Carlson's struggles during the 1930s to develop the dry photocopying process have already been outlined. The story of nylon has many similarities. In 1928 DuPont authorized one of its young recruits, Wallace H.

---

*Edison General Electric merged with the Thomson-Houston Company in 1892 to become General Electric.

Carothers, to begin work in a new field of chemistry with no particular practical objective in view. As R. Nelson (1959, p. 157) said:

> The very lack of a specific objective, the flexibility of the research project, was an important factor behind its success.

The development of polymer 66, as nylon was called initially, took more than 11 years and cost some $2 million (about $200 million in today's money) before the product was launched. Together with many other developments started in the 1930s, 1940s, and 1950s, they were backed by industrialists* who shared the vision and determination of the bright sparks who created them. The full-blooded, mutually trusting union of capital and brains is irresistible. Its harvest was the Golden Age.

As an apostle of John Stuart Mill, I believe that there is no *good* reason why the Golden Age had to end. It could have continued, and indeed, it can be resuscitated again. I believe that it was brought to an end by the growing power and influence of consensus and institutionalized thinking, processes which are inimical to sustained wealth creation. In particular, the idea that risk is synonymous with inefficiency has become difficult to dislodge. The consequences for research have been severe because Nature is characterized by uncertainty and instability. The proteins, for example, the large molecular structures that are among Nature's most fundamental building blocks, are essentially unstable. If they were not, we would not exist. Research that sets out to minimize risks is not worthy of the name. It would study a version of Nature that Nature herself might hardly recognize. Industrial thinking is also converging. In today's audited and risk-averse world new discoveries would have little chance of being developed if it was thought that their gestation might be long and uncertain, particularly if there were no clearly identified markets. Thus, we are being deprived of the modern successors to such pervasive products as the camcorder, float glass, laser, nylon, penicillin, photocopying, and transistor radio. Apparently, over the last few decades, the powers-that-be have progressively converged on the idea that Mill's vision of the "unlimited growth of mankind's power over Nature" has finally reached its limit. Should we be wondering, therefore, why economic growth is in decline?

The idea, widely prevalent today, that all the "easy" discoveries have been made is as old as the hills. It has always been proved wrong, and

*Frank Jewett, the first President of Bell Labs, said in a speech given in 1938 that the firm's research was "a currency which is more potent than gold."

indeed, it is nothing less than preposterous. It is possible that those who invented the wheel might have thought that their new world was so idyllic it could not be improved upon, and nothing so important could ever be done again. The arrival of the bow and arrow, steel, printing, and sliced bread might have induced similar musings. And so on, ad infinitum. Most, if not all, important discoveries or inventions can be made to look easy after the event, but few have been other than painfully conceived. Step changes require the creation of points of departure from what has gone before and the courage and conviction to pursue them to what their instigator believes is their logical conclusion. Indeed, it is possible that there has *never* been an easy discovery. Consider, for a moment, the world of artists or musicians. After seeing or hearing a great masterpiece, they may think that they will never be able to match such perfection, but they continue to do so. Who would dream of saying that all the easy music has already been written?

Nevertheless, many great and influential scientists venturing into the policy world seem to have little confidence in the creative abilities of future scientists. Leon Lederman (1991, p. 5) said:

Furthermore, in all areas of research, the last decade's "easy" problems have been solved, and the cost of creating new understanding of nature has increased considerably.

John Ziman (1994, p. 46) said:

[T]here can be no return to the situation where the individual scientist, aided by a few skilled technicians, could personally devise, and have constructed, the bulk of the apparatus that was needed for research.

But most of our Venture Researchers (see Chapter 8) did indeed construct key components of their equipment with remarkable results. Off-the-shelf equipment, expensive or otherwise, was simply not available. Nevertheless, such pessimistic thinking pervades the agencies that commission academic research.

Economic growth is not the only casualty when creative people are expected to march to order. The phenomenal rise in power and authority of the United States was one of the twentieth century's most stunning successes. Her experience is unique, having risen from middling status in terms of her international influence at the beginning of the twentieth century to the world's solitary and undisputed superpower at its close. Mighty nations have usually stamped their imprint on history by seizing control of as much territory as possible, but the United States has been exceptional in this respect too. Her boundaries changed only marginally during the twentieth century, and none of that was enforced. One of the less remarked of many traits the United States has made fashionable, however, is the idea that governments can focus their might on heroic problems of global proportions and bend them to their will. In the days of old-fashioned territorial conquest, victory could be expected if there was sufficient audacity, a determination to succeed, and the will to commit breath-taking levels of resources. These heroic problems could similarly be expected to fall to such relentless onslaughts. In these new respects, the United States's success has been mixed, but the legacy remains.

The first of these Herculean tasks was the Manhattan Project in which the United States, with a little help from her friends, built the first atomic bomb during the World War II. The range and magnitude of the technological problems were immense, and it was not certain* that they could be overcome on the required timescale. However, the possibility that Germany, the nation in which nuclear fission was discovered, might win the race to build a nuclear weapon could not be ruled out, and so President Roosevelt gave the atomic bomb the highest priority. See Text Box 28. The problems were solved, of course, but the financial cost to the United States was some $2 billion (1940s dollars) expended when the pressure to allocate scarce resources to other less risky projects was extreme. Indeed, in a statement issued after the dropping of the bomb on Hiroshima, Winston Churchill said that the building of the bomb:

> constitutes one of the greatest triumphs of American—or indeed human—genius of which there is record. Moreover, the decision to make these enormous expenditures upon a project which, however hopefully established by American and British research, remained nevertheless a heartshaking risk, stands to the everlasting honor of President Roosevelt and his advisers.

The next notable success began in May 1961. John F. Kennedy, in one of his first acts as U.S. president, committed his country to the

---

*The influential British physicist James Chadwick advised the Americans in 1941 that if pure uranium-235 could be made available, there was a 99% chance of it being able to produce an explosive reaction and a 90% chance of an efficient explosion.

Text Box 28: The Birth of the Bomb

The fascinating story of the discovery of nuclear fission was Europe-wide, spanned some 40 years, and involved a glittering galaxy of gifted scientists. It began with Henri Becquerel's discovery of radioactivity in 1896, progressed with Ernest Rutherford's (1871–1937) discovery of atomic structure in 1911, and ultimately culminated in experiments performed in Berlin by Otto Hahn in 1938. Although at first Hahn did not realize the significance of what he had done, it in fact turned out to be the first observation of nuclear fission. Hahn had collaborated with Lise Meitner (1878–1968) since 1917, but Meitner's Jewish origins (she had long converted to Protestantism) meant that in the summer of 1938 she was suddenly forced, in the full flow of her work with Hahn, to take refuge in Denmark to escape Nazi persecution. Only a few weeks after Hahn had written to her about his latest results, which he found puzzling because they defied conventional wisdom, she and her nephew Otto Frisch (1904–1979) correctly and brilliantly interpreted them as nuclear fission. Einstein said later that the resultant Meitner–Frisch paper published in *Nature* on February 11, 1939, was a crucial step toward the bomb. He thought it was even more important than the work of Hahn himself. Nevertheless, despite Neils Bohr's efforts to get Meitner included, Hahn was awarded an unshared Nobel Prize for Chemistry in 1944 for this discovery.

The possibility that an atomic bomb could be built was discussed in the open literature, but it was generally thought that the critical fissile mass would be measured in tons and therefore of no practical use in war. When Frisch came to England shortly after, however, he was to play a decisive role, together with Rudolf Peierls (1907–1995), in re-assessing the earlier calculations and convincing the British authorities that an atomic bomb was indeed a practicable proposition.

Apollo Project, which entailed sending a manned spacecraft to the Moon "... before this decade is out." The necessary technology was already in place, in principle, in 1961—hence Kennedy's confident assertion. The Americans had already put men into Earth orbit, but a mission to set foot on another planet is vastly more difficult. Technological demands—in terms of reliability, performance, and accuracy—would be unprecedented, and it was by no means certain that they would be achieved. Personal and organizational skills would be similarly stretched. Unlike the Manhattan Project, however, every step was illuminated by the full glare of global publicity. It was, of course, a consummate success and a pinnacle in humanity's achievements. The American government and the National Aeronautics and Space Administration (NASA), in particular, deservedly won the applause of an inspired world.

Perhaps President Richard Nixon was also inspired when he launched "a war on cancer." In 1971, he signed the National Cancer Act, which gave the National Cancer Institute (NCI) special status within the National Institutes of Health. The NCI has spent more than $30 billion since that time trying to unravel the basic biology of the disease and to develop new tests and therapies. Much has been achieved in the understanding of cell biology, but clinical success has been modest. Mortality rates from some cancers have declined while others have increased.

In 1984, President Ronald Reagan set up the grandiose Strategic Defense Initiative (SDI) to shield the United States from a possible Soviet missile attack. The SDI was one of the most contentious projects ever launched by a government. Umbrellas are useless, of course, unless they are completely waterproof. For SDI, if the U.S. defenses were penetrated by only one missile from the hail expected in a full-scale attack, it could mean failure. Since neither detection nor destruction efficiencies can be 100%, the project was doomed from the outset, as many Americans passionately argued. Some $33 billion and nine years later, the authorities agreed. The project was then renamed the Ballistic Missile Defense Organization (BMDO) and was given the primary task of defending battle areas against ballistic missiles. (The BMDO was replaced in 2002 by the Missile Defense Agency with the primary task of defending the United States, deployed forces, allies, and friends from ballistic missile attack.) However, the SDI also triggered an arms race. Its astronomical cost seems to have been a major factor in bringing about the Soviet Union's subsequent downfall. One could argue, therefore, that President Reagan's peace shield was successful!

In 1990, U.S. government institutions took the lead in starting the Human Genome Project in collaboration with the United Kingdom, France, Germany, and Japan. It has now achieved its goal of sequencing an entire human genome, a *tour de force* and one of humanity's greatest achievements, at a cost approved by the U.S. Congress of $3 billion. It has raised many thorny issues, however, such as patent rights* and how limits might be set on the commercial exploitation of the vast quantity of data. See Text Box 29. Biology is of course a vast and complicated

*Irwin Chargaff, a pioneer in the field of genetics, has in his characteristically robust style described the patenting of human genes as "a molecular Auschwitz, in which genes are extracted rather than gold teeth."

Text Box 29: Junk

The successful sequencing of the human genome and other related projects now means that scientists are awash with data. Merely writing down the linear sequence of the human genome would fill a 10-million-page book with a largely unpunctuated string of letters! However, mountains of facts are not the same as understanding. We still understand very little, and scientists are often denied the time for quiet contemplation in the rush to publish and protect their careers. The genome, for example, is a three-dimensional time-varying structure—that is, each gene has a three-dimensional "address" within the genome that determines the local genetic environment in which the specific gene must function. The environment can be crucial to a gene's operation. Move it to what turns out to be an inhospitable location and it may stop working even though the gene itself might not have changed. Unfortunately, most of today's work on genetics focuses on linear sequences. Another area of uncertainty, and one of the most profound mysteries in biology, is that the active genetic sequences in plants and animals are only a small part of the whole genome. For humans, it is only 3%. That is, 97% of the human genome has no known function: The blind man exploring the proverbial elephant has contact only with its trunk! Being unknown, the 97% was often called "junk"—the clear implication being that our lack of understanding is not really important. Yet, every time a living cell divides, the entire genome—junk and all—is faithfully duplicated. It would indeed be astonishing if Nature were to expend so much energy unnecessarily. In 2003, it was gradually being recognized that junk might contain a multitude of hidden and unsuspected controls vital to life. I hope it gets the attention it deserves.

Pat Heslop-Harrison, a Venture Researcher (see Chapter 8), has proposed an interesting thought experiment. If the known active genes in one plant species, say, rye, could all be removed and used to replace the active genes in another plant species, say, barley, what would be the result? The junk in each species would stay behind, but what would its effects be? See Figure 13.

field. Intractable problems abound, and understanding is patchy, to say the least. However, I never met a biologist who thought that sequencing genomes should be given the highest priority, as the level of funding for the project implies. It seems to have been launched *because the goal was*

Figure 13
Two Venture Researchers (see Chapter 8)—Mike Bennett, Jodrell Laboratory,
Kew Gardens, and Pat Heslop-Harrison, University of Leicester—working to
show that the chromosomes in the cellular nucleus are organized hierarchically
and that a chromosome's position influences its behavior. Before their work, it
was generally believed that the chromosomes swam in a nuclear "soup" or, to
use an alternative analogy, had a similar structure to a dish of spaghetti.
(Photograph by BP in 1989 and reproduced with permission.)

technologically accessible. Unfortunately, the funding of other less accessible problems of more pressing urgency has suffered. In addition, the
demand for expensive equipment has soared. Unless individual scientists
have access to the latest and most efficient sequencing technology, they
cannot compete.

There is a common factor in all of this. When nations declare hostilities, so to speak, against formidable problems, one hopes they will be successful. But that can only happen if one or more of Nature's strongholds that hitherto have defied all attempts at penetration can be breached. It is more easily said than done.

As Francis Bacon so presciently pointed out long ago, "Nature is only to be commanded by obeying her." Nature's ways are, of course, extremely complex, and we understand very few of them. I believe Bacon's apparently cryptic use of the word "obeying" is in the sense of "submitting to the authority of." He would seem to be implying, therefore, that any discoveries we might make would be made only with her permission, so to speak, because we have somehow deserved the right to be successful. Thus, I believe that Bacon's warning is that if we set out to conquer one of Nature's domains, we should be careful to choose one that Nature might be prepared to yield. We have long known that Nature is undemocratic, which implies that she might also be immune to eyeball-to-eyeball confrontation. We should expect, therefore, that throwing virtually unlimited amounts of money or resources at intractable problems would probably be unsuccessful *unless the ground has been carefully prepared beforehand.* This means that when a powerful organization announces that it intends to bring one of Nature's untamed dominions under control by a certain date, it should at least have consulted her about what it has in mind and listened very carefully to what she has to say. To put it less fancifully, it means that it should have instigated extensive research that hopefully has asked the right questions and thought deeply about the results. Unless the ground has been sensitively surveyed, we should not be surprised if our expensive commitment is not rewarded by the anticipated success.

In the case of the atomic bomb, the ground had been well prepared. Two years of inspired work by many British-based scientists resulted in the dramatic conclusion that the bomb was a possibility, but there was absolutely no way it could be built in wartime Britain. Such an enormous project could not have been kept secret on a small island. It would also have been vulnerable to bombing, and furthermore, wartime Britain simply could not afford it. There was no alternative but to turn to the United States, whose response was heroic. Although the principles of the problem had been broadly worked out, solutions in practice are rarely achieved without considerable ingenuity. The sheer scale of the U.S. effort too was formidable. For Apollo, NASA's research and development over more than a decade had indicated that a manned expedition to the moon was on. But President Kennedy's panache and flair, his courage, the resources he could authorize, and his confidence in NASA's abilities were needed for his declaration to the world that it could and would be done.

For cancer and other major diseases, however, Nature was and still is being ignored.* At the most basic level, we know a great deal about the chemical reactions that govern the operation of a living cell. But we understand little about the ways that even two healthy cells communicate and how information from the many different pathways is integrated and regulated to produce healthy tissue. If we do not understand one-to-one cellular communication, how can we expect to put it right if it goes wrong for many millions of them? For AIDS, it is generally believed that the disease is caused by the human immunodeficiency virus (HIV). However, viruses are perhaps the least understood of all living† organisms. In particular, the phenomenon of latency, by which a virus can lie dormant within a host for long periods, simply ticking over without either attacking or being attacked, is hardly understood at all.

Is any of this relevant to today's global problems? Unfortunately, it is. Their very scale and importance can make them tempting targets for governments and other powerful agencies that might want to stamp their authority on the world. Crusades tend to be popular. There would be no shortage of scientists willing to take part in such direct onslaughts because at the very least they can be sure of sustained funding. The media too would usually be supportive. "It is about time that something was done about disease, or global warming, or population, or pollution, or whatever." Nevertheless, Nature is unlikely to be impressed by such combative assertions. See Text Box 30. She is never devious, but if our frontal attack is based on misconceptions or poor intelligence, Nature's answers to our wrong questions will probably be unhelpful. They might even be misleading. After countless billions spent on AIDS research, there have been no reports of a person being cured. Indeed, if the AIDS pandemic proceeds at its present rate, some 45 million new infections are expected by 2010, with 70 million deaths by 2020. The cost per person of containing HIV infections in a rich country can be much more than 10 times the annual income of victims in the poorest. The disappearing rain

---

*There have been some important successes: smallpox, for example, would seem to have been eradicated from the world. This important goal was achieved progressively, however, after many years of patient research, as our understanding of the disease improved. Its eradication was not initially "declared" a global objective.

†Strictly speaking, a virus is not a free-standing organism but requires a living host in which to reproduce. Viruses have aptly been called "bad news wrapped in protein."

Text Box 30: The Ozone Hole

Sir George Porter (later, Lord Porter, 1920–2002) once remarked when he was President of the Royal Society, "who would have thought that the study of an obscure and transient chlorine molecule would lead some thirty years later to an accurate diagnosis of the Ozone Hole problem?" The molecule was chlorine oxide (ClO), which has a lifetime of about a millisecond. His work on ClO was part of a series of experiments that led him to win the Nobel Prize for Chemistry in 1967. He made the remark in a British national television program dedicated to Venture Research—*Blue Skies* (see Chapter 8). Porter was a keen sailor. He spent some of his prize money on a boat, which he called *Annobel.*

forests may be contributing to global warming. However, overlogging seems not to be the problem. Forests are being lost to agricultural encroachment, industrial pollution, and a host of other nonforestry factors.

But the questions we should be asking are not always obvious. They usually need the most careful consideration of every conceivable aspect of the particular problem and, more often than not, an inspired genius. Sometimes, the *inconceivable* aspects turn out to be the most important. However, if we want to encourage thoughtful reflection, we need to provide the free and flexible environments in which it can take place, and they are thin on the ground nowadays. Those who control the purse strings will rarely authorize excursions outside what they consider to be the main parameters of the problem.* The danger is, therefore, that the high-profile attacks authority launches from time to time will soak up most of the available resources. Not much will be left, therefore, for such playful activities as those that have led to the laser and other unpredictable discoveries. But vital clues to our major problems might well emerge from eclectic combinations of these eye-opening revelations. See Text Box 31.

As ever, the resuscitation of creativity will be a major factor in our

*Most of the major advances in biomedical research, for example, have come from such "external" fields as physics and chemistry. Nevertheless, biomedical funding agencies will rarely directly support research in these latter fields. They seem to have forgotten Pasteur's famous observation that there is only one science.

Text Box 31: Jumping Genes

The American scientist Barbara McClintock (1902–1992) shared with Marie Curie (1867–1934) the distinction of being the only woman in the twentieth century to win an unshared Nobel Prize. Curie's story is legendary, but McClintock is not as well known as she should be. After many years of research using maize as a model system, a radical innovation in itself, she discovered during the 1950s that some genes can control the expression of other genes, and some can vary their position on the chromosome—widely known as jumping genes or more formally as transposable elements. At this time, it was generally thought that a gene's location on the chromosome was fixed, as was its function. Not surprisingly, therefore, her discoveries were initially greeted with silence or disdain and, since they had been made on such an unfashionable crop as maize, could be dismissed as an exceptional curiosity. However, they proved to have immense significance. In particular, they played a major role in understanding the mechanisms of disease resistance. Immune systems are extraordinary in that they can produce antibodies to diseases (or toxins) to which they have never before been exposed. If genes were indeed fixed, as orthodoxy required, there would need to be perhaps a million genes expressing antibodies (which are proteins) if the immune system were to protect against all possible invading antigens. That was known to be impossible. However, her seminal work inspired Susumu Tonegawa, a Japanese scientist working at the Massachusetts Institute of Technology. He transformed our understanding of the immune system's development and refined the concept of individuality among living cells. Although the immune system's cells all have identical structures, individual cells can shuffle some components of the "gene pack" at random, so that each cell can assemble a slightly different gene. Thus, the millions of cells of the immune system are able to synthesize millions of different antibody-producing genes from the shuffling mechanism, and the healthy body is ready to tackle virtually any alien invasion, hopefully on demand.

McClintock had to wait some 20 years before the significance of her work was recognized. Thankfully, she lived long enough. She won the Nobel Prize in 1983, as did Tonegawa in 1987.

survival. As has been argued in Chapter 6, however, funding is almost invariably tied to specific objectives, and restrictions usually inhibit creativity. As I have said before, but which can hardly be repeated too often, this situation is new. Nowadays, the geese that would lay the golden eggs are increasingly being farmed and focused. In particular, they are being urged to lay only those eggs whose shape, size, weight, and luster conform to the required specifications. It is no wonder, in these circumstances, that our magical geese have become egg bound in recent times.

Not everyone agrees that economic growth should be an imperative. The Greens and other environmental activists argue that we should try to live in harmony with the world as it is. We should be content with what we have. This pastoral policy is superficially attractive, but I do not believe it is a viable alternative. The Orwellian scenario briefly outlined in the last chapter could become a nightmare if global per-capita economic growth continues to fall or hovers dangerously close to the population growth rate. We must have growth because we do not know how to run economies or organizations that are not expanding so as to maintain a reasonable quality of life, at least for the majority. Global growth has increased inexorably since the Industrial Revolution started more than 200 years ago, so we have not had this problem for a long time. But before that, we often had to endure extended periods of stagnation or falling growth, when life was miserable for almost everyone. Most of us today seem to want opportunities for advancement or more leisure or more money, and they will be few and far between when one's organization is stuck at the same level year after year or, worse still, running down. All this seems to be true at any institutional scale from small laboratories to companies or nations. Thus, for well over 100 years, we in the industrialized countries have become accustomed to expect *rising* levels of production or consumption. The Greens never seem to address these issues.

On the global scale, it is an obvious fact that if average economic growth is close to zero, some nations will be better off in the future, but if they are, others must stay the same or fall back. The unlucky ones will therefore be constantly beset by the crises of everyday survival. Most of the increase in world population comes from the developing countries. If global growth continues to decline, political pressures in the advanced countries could jeopardize even the small portions of GDP transferred overseas through aid programs or cancellation of debt. Charity begins at home. On the other side of the coin, it is unrealistic to expect the poorest nations to acquiesce to rising levels of deprivation as their populations increase. They would want to do something about it. All too often, they will have that option, as cutting expenditure on armaments is often the

last economy needy countries will accept. We live in a highly intercon-
nected world. The poor today are constantly reminded of their lot by
tourism, advertisements, newspapers, television, films, and other trap-
pings of civilization. Sustained deprivation could have widespread and
unpredictable consequences, and increased risks of terrorism, conflict,
and civil war are obviously among them. Civil war is a major killer in the
developing world, and the number of wars has increased steadily since
about 1960, which is the approximate time that global GDP per capita
began its current decline. Indeed, an exhaustive study by Collier and
Hoeffler (2003) of 161 countries between 1960 and 1999 found that for
each percentage point a country's economic growth rate rises the risk of
war falls by a percentage point. Protracted periods of falling or low levels
of global per-capita growth would lead us, therefore, to a hazardous
future in which even the advanced countries could have little control. Instability and unrest could become the norm everywhere.

Wealth distribution has always been a political issue, but if global
growth is increasing, the have-nots stand a better chance of having more,
albeit marginally, through such means as technological catch-up, or
trickle-down. *The Economist,* in a special survey of technology and development (November 10, 2001, p. 3) pointed out that higher economic growth
rates improve the lot of the poor as well as the rich. The article refers to
the World Bank finding that technical progress was the biggest single
cause of reductions in mortality between 1960 and 1990, highlighting
such discoveries as antibiotics, vaccines, and oral rehydration therapy.
Impeding the creation of new and appropriate technologies, whoever the
intended beneficiaries might be (remember Adam Smith's invisible hand),
might sentence millions to a miserable existence or death. Growth creates
flexibility and new options. Stagnation imposes a straightjacket.

However, there can be dangers when new products or processes with
high economic value are created without rigorous attention to the underpinning sciences. Once upon a time, any new technologies humanity was
lucky enough to come across came long before we fully understood how
they worked. Trial and error had to be accepted because there was no
other way. But we should know much better now. The biggest concerns
seem to be in biotechnology. In the controversial case of genetically
modified foods, genes from, say, bacteria have been transferred to plants
to give them resistance to insects or from Antarctic fish to give tomatoes
resistance to frost. There are many other examples. The general theme is
that desirable genes from one organism can be successfully transferred to
another. One must hope that the gene's performance will not be affected
by what is for the gene a radical change. One must resort to hope because, although the technology that enables such transfers to be made has

been refined and developed over many years, progress in the corresponding basic sciences has not kept pace. For one thing, it is much more difficult to get support for them. This is especially true if the proposed research were not tied to a particular priority or if their objective is "merely" to advance knowledge because the researcher thinks the problem is fascinating. One usually has to be specific about *who* the beneficiaries will be for a proposed advance before support will be considered. Responding with the word *humanity*, as one should be able to, would not be enough. This means that one's chances of getting support are much higher if one's basic research is targeted. But Nature rarely responds to confrontation, as has been said many times. Subtlety is much more effective.

In biotechnology, therefore, technical expertise has outrun basic understanding. The genome of any living organism is astronomically complex. The intimate relationship between the genome in the cell nucleus and other cellular components such as the mitochondria are usually ignored. We do not know, for example, why some genes fail to function when inserted into the genome of another organism. This also means that we do not know why some other genes may, apparently, be successfully transplanted. It is a precarious situation. This does not mean that we should wring our hands and do nothing. If only we would commit a tiny fraction of the resources spent on R&D generally to purely impartial research, we would soon have solutions to problems we did not know we had. We would also have a better understanding of the limits of what can be done with reasonable confidence and what (presently) cannot. Venture Research (see Chapter 8) is one way of making such unpredictable progress.

However, there is always a downside to change, and one hopes that the net gains will be positive. But it is easy to bias one's focus. Environmentalists do us all a service when they expose those companies and organizations that pollute and contaminate. Apparently, however, environmentalists do not seem to realize that such wantonness may sometimes stem from ignorance and from the expediency that comes from simply trying to cope. See Text Box 32. Commercial pressures can lead to corner cutting, but once again, some of the ensuing problems can usually be traced to ignorance. We do not always know which corners can be safely cut.

But it must be acknowledged that some companies and organizations are driven by rapacity and greed and are totally indifferent to the damage they do or the suffering they cause. Such people seriously threaten our environment, indeed our very society. We all have a duty to do what we can to stop them. Some people will abuse any system of governance, of course, but that does not necessarily mean that the system is flawed. In

---

Text Box 32: The Tragedy at Minamata Bay

Minamata Bay paralysis was first noticed in the mid-1950s in the Japanese villages surrounding the bay. Residents experienced paralysis of their hands and feet, pain, chills, headaches, and impaired vision and speech. These symptoms later developed into mental disorders and many died. Survivors suffered from incurable paralysis and pain, while large numbers of women either miscarried or gave birth to deformed children. It was eventually discovered that the disease was caused by eating fish contaminated with methyl mercury coming from the nearby chemical plant discharging waste into the bay. At that time, there were few government controls on pollution in Japan, as production was seen as an imperative. However, even if controls had been in place, there may still have been problems. When the causes were finally established, it was found that the company had been discharging mercuric chloride, which has low solubility and is relatively easy for the human body to expel if it is ingested. Tragically, however, and unknown at the time, marine microorganisms convert this relatively harmless pollutant into the highly toxic methyl mercury, which then enters the food chain. This awful story took many years to unfold. The bay was opened to fishing again only in 1997, and the compensation claims settled only two years before that—an appalling delay.

---

this case, the "system" in question is the concept of wealth creation, and to seek to abolish it, or severely to constrain it, because some people are wicked makes no sense at all. Even if acceptable alternatives to wealth creation could be found and we could live more or less tolerably in economic equilibrium, some people would abuse that system too.

The state of equilibrium or lack of change, however, is rarely found in Nature. The most stable system we know is the proton, the nucleus of the hydrogen atom. Its life has been deduced to be much greater than 10^{30} years—the upper limit is not known. Nevertheless, we do know that the proton, *this most stable of entities*, has on close inspection a complex structure that commands the attention of thousands of researchers. The so-called vacuum state, the name given to a volume of space that is apparently devoid of stable matter, is a seething state of restless uncertainty when probed over sufficiently small timescales and distances. These quantum fluctuations, as they are called, are unavoidable. The vacuum state exists everywhere, and so, of course, it is included within the living tissues of every one of us. Our planet Earth has been around for almost 5 billion years, but every aspect of the land, sea, and air environments, like the weather, has never ceased changing. It is possible, therefore, that we may derive new ways of civilized life that to a remote outside observer may appear for long periods to be static and unchanging. This may be what the Greens and others have in mind. We have no idea how such a society might be constructed, but it does not seem impossible. However, we are products of Nature, and the internal structures of that idealized society, and the lives of its people, would be constantly changing. Most definitely, they would not be stagnant. We would have learned how to live with change.

Societies that tolerate dissent are diverse. A thousand blossoms bloom, optimism abounds, and healthy growth usually follows. On the other hand, societies that suppress dissent tend to be monolithic, as they were in the stagnant Dark Ages and are now in totalitarian regimes. One cannot be optimistic in a monolithic society because there is little scope for change. The short term and the status quo are all there is. Wealth is difficult to generate in such societies. The few who are born with it keep it for life. Those who would oppose the *concept* of wealth creation should realize, therefore, that they would wish to curtail diversity and suppress dissent. I doubt if most of them would wish to live in such a world. Wealth creation can lead to conspicuous excess, of course. Exploitation and corruption can be part of the downside. But the best antidote to the sins of omission or commission is to generate more understanding. Environmentalists who take a blanket antiscientific stance should recognize that they could do incalculable harm. Science and new understanding are the only lifelines that can stop us from being sucked into the pools of stagnation.

However, some scientists would place some limits on exploration. Most notably the very distinguished British scientist Sir Martin Rees (2003) describes in chilling detail some of the risks humanity faces unless scientific progress is slowed down. Rees has in mind such nightmare scenarios as nuclear terrorism, experiments that might have such unintended consequences as creating mini–black holes, biological and other experiments that could go seriously wrong, or the deliberate misuse of scientific knowledge in general, any of which could seriously threaten or wipe out humanity. Furthermore, one of his main themes is that "technical advances will in themselves render society more vulnerable to disruption." Indeed, he estimates that the human race has only a 50:50 chance of surviving the present century. The time has come, therefore, to curb scientific exploration, even "pure" research, if there is reason to expect that the outcome will be misused.

Rees's scenarios are not fanciful. They could happen. However, I believe that his justifiable concern stems from the scientific enterprise's current obsession with outcomes, with specific technologies or results deemed relevant to some collectively imposed objective. Scientists are losing the freedom to be impartial. They have been obliged to become eager to please, to satisfy their paymasters, rather than, as they should be doing, satisfying their own curiosity. If the enterprise is dominated by the need to bring forward imagined worlds based on extrapolations of current knowledge, nasty people can easily jump on that bandwagon. However, if it were concerned primarily with creating new understanding—*pure and disinterested understanding whatever its purpose*—such unpredicted results as X-rays, lasers, or computers might astonish any imagination. Its purposes would then be much more difficult to corrupt. But any knowledge can be misused. Quantum mechanics could be said to have led us to atomic weapons and computers that could control the world, and molecular biology to an all-pervasive "grey goo" that could devastate life on Earth. Should these and other originally curiosity-driven enquiries have been stopped?

Vigilance should not imply proscription, and I believe that the last thing we should do is to declare some areas "no go." The scientific enterprise is essentially intellectual, and as the Inquisition, the Puritans, the Soviet Union, and many others have discovered throughout the ages, attempts to control or direct what people should think are eventually doomed to fail. In any case, the proscription of some types of knowledge, or even decisions to slow down progress in some areas, when coupled with the routine effects of bureaucracy and control discussed in this book would compound our problems. We need to expand our horizons, not contract them. If we do not, it may well be our final century, at least for civilized life as we enjoy it today.

None of this should be taken to mean that the global economy or its population could be envisioned to grow without limit, of course. But we understand so little that it can be safely assumed that those limits are nowhere near in sight. Growth in the economy must keep ahead of growth in population, but that does not seem to be an insoluble problem. We have a breathing space, therefore. The more we promote science and new understanding, the more we prolong civilization.

Nowadays, scientific play is deemed wasteful and an inefficient use of scarce resources. Scientists, however, have never had enough money. *They never will.* New uses for money can always be found, but they are not necessarily useful. Ernest Rutherford's famous dictum was "When money is short there is no alternative but to think." In the next and final chapter, I shall try to explain one of the advantages of persisting with that painful process.

Venture Research

In the days when armies laid siege to fortified towns, a common practice was to force a breach in the weakest section of the walls using the heaviest artillery at hand. After a few days intense bombardment, even the thickest walls would crumble. A small detachment of volunteers could then be sent in to secure the breach in advance of the full-scale attack. The snag with this strategy was that the defenders had precise warning of where and when the attack would come. The detachment's reception would therefore be hostile in the extreme, and few could expect to survive—hence the call for volunteers. Amazingly, perhaps because of the euphoria of the heat of battle, volunteers were readily available. There was also the (doubtful) incentive that those who survived were promised a glorious future and instant promotion. They would also probably enjoy the acclaim of their fellows. The Dutch name for these almost suicidal detachments was *verloren troop* ("the lost troop"). To the British, with their flair for anglicizing the sound of foreign words, it became the *forlorn hope*.

The origins and subsequent fate of what came to be called the Venture Research initiative has much of the forlorn hope about it, although the attack, so to speak, is still in progress. I might not survive, but the

Pioneering Research: A Risk Worth Taking, By Donald W. Braben
ISBN 0-471-48852-6 © 2004 John Wiley & Sons, Inc.

prize is so great that others may be tempted to volunteer. This outline is intended to help them avoid some of the pitfalls, the worst of which may come as a surprise. The battle to persuade Nature to reveal her secrets is, of course, mainly intellectual. Nature's subtleties are rarely uncovered by naive probing, but a successful interpretation of what seems to be a mystery might not always be met with acclaim. One's peers have also to be convinced, and if vested interests are involved, their opposition can sometimes be less than subtle. If one's new perspective would radically challenge convention, it may seem sensible therefore to prepare for a hostile reception from one's colleagues. That is often unnecessary. The frustrating fact is that he or she is likely to be ignored.

I discussed some of the Venture Research story in Braben (1994). My purpose then was to describe how one might become a scientist, how to choose a problem, the differences between research in the natural sciences and engineering, and how one can get published or win prizes. The politics of science and other extraneous issues were only touched upon. This is not the place for a full history, but I will expand the story somewhat in the hope that the experience may be useful to those who may wish to launch their own crusades.

When my secondment to the Cabinet Office ended, I returned to the Science Research Council (SRC), as it was called then, and was given responsibility for two initiatives. The first was the Marine Technology Directorate, set up on the threshold of the United Kingdom's exploitation of its newly discovered offshore oil and gas deposits. At that time, the SRC's annual expenditure on marine technology research was only some £0.5 million. This hardly reflected the growing importance of the field to the United Kingdom. It arose because SRC's committee structure had a roughly one-to-one relationship with the academic disciplines while marine technology embraced many of them. A properly constructed study of corrosion in a marine environment, for instance, might need input from, say, civil and mechanical engineering, physics, chemistry, and marine biology. Academics who submitted coherent proposals in this new field would find that they were bounced from committee to committee. No single committee would want to support work that could be argued was the responsibility of another, especially when budgets were tight. Academics are very sensitive to this sort of message, and the SRC's paucity of involvement was the result.

The SRC decided therefore to make special provision for research in the new discipline and set up a Marine Technology Task Force to derive a possible program. Jack Birks, a BP managing director, was invited to chair it, and its membership included other senior industrialists and academics with a wide range of backgrounds. The report* we produced described the urgent need for the academic sector to build up its expert knowledge base and to develop its research in a number of key areas. Training was also covered. The SRC approved all of the task force's recommendations and a five-year program costing some £25 million.

My other responsibility was the Teaching Company. This SRC initiative was partly based on ideas behind a teaching hospital and was designed to bring academic engineers into closer contact with industry. Thus, newly trained engineers were encouraged to tackle real industrial problems, and while they would have some academic supervision, they would spend most of their time in an industrial environment. It was not essential that the work should lead to a higher academic qualification. As "officers" of the council—an impersonal title I always hated—our main job in this pioneering initiative was to find companies willing to participate in the program. It was not easy. Mutual misunderstanding and mistrust between academic and industrial engineers are probably higher than for any other profession. We also helped to develop the agreed-upon work and tackled such problems as confidentiality and patent rights.

We also needed to recruit senior professional engineers to head these directorates, and my contacts with headhunting firms led to being headhunted myself. MSL, then a large London recruitment firm, telephoned to ask if I would like to be the "chief scientist" at the Bank of England Printing Works. He said it was a very high tech job. The bank printed millions of bank notes a day, and it seemed that they wanted an adviser on printing technology. "But I know absolutely *nothing* about printing," I protested, to which he replied, "You are precisely the sort of person they are looking for! They want someone who will introduce new ideas."

Some 15 hours of interviews later, I duly accepted the prestigious job, but it did not take long to realize that it was not the wisest of decisions. Bank-note design and production are indeed complex processes, but the pace of change is glacial. The production of a new bank-note design, for example, usually takes years. The adviser's role can be important, but once the advice has been given, the responsibility for any subsequent action passes to the executives. The adviser must then wait passively for the next consultation. With time on my hands, I had no alternative but to think and naturally turned to the state of the enterprise in which I had spent my career.

As luck would have it, one of the external governors of the Bank was Sir David Steel, who was then the chairman of BP. He was also a mem-

Marine Technology, Science Research Council, 1976.

ber of the Printing Works' governing board (the Debden Committee), and so our paths often crossed. He must have noticed that I was not entirely happy with my lot because his BP colleague Jack Birks phoned to ask if I would like to play a part in their new Blue Skies Research initiative. Would I indeed! BP was Britain's largest company, and I had long admired its flair and precision. The building of the Trans-Alaska pipeline in which BP had played a major role had just been completed. It was one of the engineering miracles of the twentieth century. It transported oil from Prudhoe Bay in the Beaufort Sea 800 miles overland to the port of Valdez in the Gulf of Alaska. Its 48-inch-diameter pipe traversed tundra, forest, muskeg, snow-covered and glaciated mountain ranges, frozen silt and bare rock, caribou migratory routes (for which the pipe had to be buried), and some 600 streams and rivers, including the Yukon. A small section of it is shown in Figure 14. It had cost some $7.7 billion, at that time the most expensive civil engineering project in history. A company that could do all that must have vision, and so the offer was irresistible.

Birks was then managing director responsible for almost all of BP's technical interest—exploration, production, engineering, research, and development. The company was "cash rich," as were almost all the oil majors at that time (1979), and BP wanted to invest in the research that might lead to new interests outside its current businesses. BP had set up a committee to examine the potential of Blue Skies activities about a year earlier, but it had not come up with anything that was not being done already. Would I like to write a report? The Bank agreed that I could, Birks approved it, and I joined BP in 1980 to implement it as head of the newly created, prosaically named, Venture Research Unit (VRU).

In those far-off days, managing directors of large corporations were usually free to take decisions as they saw fit. As far as I know, Birks consulted only one person about the appointment—Robert Belgrave—a classicist and head of corporate planning. My only interviews were with Belgrave so that he could explain the terms and conditions for the appointment, which was for two years initially. Most conspicuously, Birks did not, apparently, consult the director of research,* Professor John Cadogan. As I had recommended in my report, it was accepted that I should not be located within any of BP's research establishments. I feared that this might create tensions, and I did what I could to minimize them. While I needed to keep abreast of BP's interests, of course, my responsibilities should lie entirely outside them. If my thinking merely extended

*Cadogan did not formally receive that title until 1982 when Birks's attempts to recruit "a heavyweight," in his words, for board appointment fell through.

Figure 14
A small section of the 48-inch-diameter Trans-Alaska pipeline. (Photograph by
BP. Reproduced by permission.)

their's, I would in effect be competing, and I would fail. We had a difficult
tightrope to walk, therefore. We had to choose areas that had no overlap
with BP's current activities, but they would eventually have to be of some
interest to industrialists. BP is a very large company, and so another major issue was the line of command. I was to report to Belgrave on all administrative matters and directly to Birks on all others. This was a perfect
arrangement to my mind. If we were to create new business options,
where else should they first be considered but corporate planning? I
should also admit that Belgrave's Corporate Planning Unit was BP's
"Cabinet Office," and I instantly felt at home there.

The biggest issue of all, however, was what to do with the freedom I had been given. People are the key to all research, of course, but which people and where to look for them? The fashion then as now was to identify the fields expected to grow and, through consultations with the great and good, to list the most important problems to be solved in those fields. That indeed is what I had helped to do with marine technology. But someone else defined that job, whereas now I was free to do anything that BP could be persuaded to support.

BP's interests were extensive, and overlap would not be easy to avoid. But I had given myself a much bigger problem in that I did not want to compete with *any* company or organization. The research we would support had to be unique, therefore, which in turn meant that we should not allow our search to be restricted by artificial boundaries. Unfortunately, we live in a well-fenced world. Companies are set up to exploit specific business opportunities and must account to their shareholders if they stray from them. Universities and colleges are structured around the disciplines, and the boundaries can sometimes be impenetrable. If we were to be free from all these constraints, the problem would chiefly be intellectual. The old infrastructure had to be abandoned, but what could we use in its place? As my thinking was at its woolliest at that time, I visited hundreds of scientists in Britain over the following couple of years to discuss the problems with them. See Text Box 33. Most tried to persuade

---

Text Box 33: Freedom

My consultative travels led me to a very eminent physicist who told me that complete freedom would paralyze the imagination. Scientists, like musicians, he said, must have a structure within which to work. That's true, I replied, but there are many types of music—Western, Arabic, Indian—whereas there is only one science. In any case, any structures must not be externally imposed. We could not agree. The problem, of course, is that any structures that Nature recognizes are most unlikely to be the same as those adopted by universities or committees worldwide. In the very old days—in Isaac Newton's time, for example—universities recognized only one science—an amalgam of natural philosophy and mathematics. Today's departmental structures have slowly and arbitrarily evolved so that teaching and research might most easily be administered. They tend to be rigid. It can sometimes be easier to collaborate with someone in a different country than with a colleague in another department at the same university.

me to support their pet projects, but a surprising number rose to the occasion. I could hardly have given them a bigger challenge. In effect, I was asking how research should best be selected and supported if all bets were on and we were starting absolutely from scratch. The most memorable advice came from Sir Geoffrey Wilkinson, a Yorkshireman, a Nobel Laureate in Chemistry, and a professor at Imperial College London. After discussing my dilemma, he told me, with his typical north-country directness, that I should not to listen to the usual fashionable claptrap that was doing the rounds. I should do something original.

I was not entirely free, of course. All VRU's research expenditure had to be authorized by the Venture Research Advisory Council (VRAC), one of the most prestigious committees in the company. Its chairman was Sir James Menter, Principal of Queen Mary College London, a nonexecutive BP director, and a distinguished physicist and industrial scientist. Sir Hans Kornberg, Professor of Biochemistry at Cambridge (later to become Master of Christ's College Cambridge), and Sir Rex Richards, Warden of Merton College and Professor of Physical Chemistry at Oxford, were also members. Ex officio members from BP were Professor John Cadogan as Director of Research; the head of corporate planning, Robert Belgrave initially; and the chief engineer.

Oscar Roith was the first chief engineer to serve. He left BP later to become the chief engineer and scientist at the Department of Trade and Industry in Whitehall. However, Roith liked to describe himself as the peasant on the committee, such was the elevated status of its other members. Indeed, the VRAC was perhaps only slightly less glittering than BP's main board, an august body that oversaw the company fortunes in general, with annual revenues of some £26 billion in 1980. VRU's total budget never exceeded £3 million pa—the equivalent of about an hour of BP's turnover—and amounted to some £15 million over the decade of its existence. In big companies and elsewhere, perceptions of importance are usually related to cost. In giving prominence to an activity with relatively trivial outgoings, BP had recognized, therefore, that the numbers game is not necessarily important in research. One person can succeed where thousands have failed. Little acorns can reliably grow into mighty oaks if care is taken to pick only the best.

Formidable intellects are not easily swayed, and so just how free would I be? I told myself that the VRAC's daunting power might not necessarily be bad. If I could persuade them to do things differently, it could mean that they had accepted the new methods, and I would have powerful allies. That did not turn out to be wholly accurate, however. All the VRAC scientists were Fellows of the Royal Society, which can be a rather conservative body. While they were often willing to accept my recommendations for the support of a particular scientist, they seemed unwilling to agree that the conventional selection procedures had failed or even that they were capable of failure if the proposed science was any good. They seemed to think that I had merely been astute enough to sign the scientist up for BP so we should have an inside track later if the research was successful. Their philosophy, whatever it really was, was a matter for them, of course. We had to ensure that we had the most objective selection criteria that could be derived. My problem thereafter was to convince the VRAC that *all* the scientists who met those exacting criteria should be supported, but I soon realized that it would have to be done on a case-by-case basis.

Of course, we would stand or fall on those selection criteria. If our scheme were to be so original that it would have no competitors, we would need to strive for absolute standards of excellence. That precise goal is probably unattainable. However, we would try to make our selections as objective as possible, that is, as free from bias and prejudice as we could make them. The scientists we sought would all be pioneers, and pioneers usually need total freedom. How could they be reliably identified? Luis Alvarez, one of the most gifted American physicists of modern times, once said that the peer review system in which proposals rather than proposers are reviewed was the greatest disaster to be visited upon the scientific community in the twentieth century. As Alvarez's opinion resonated with my own, our first step would be the abolition of peer review. This was perhaps our most difficult decision. It was also dangerous, as peer review enjoys a state of grace among scientists today that approaches the sanctity of motherhood. We in the VRU were necessarily engaged in selection, of course, but we were not in competition with the applicants for funds. As we were not competing, we would not be putting our reputations in jeopardy if a proposed new line of enquiry were deemed by conventional wisdom to be either impossible or irrelevant. Nor would we have egg on our faces if it succeeded. But as acolytes of Alvarez, we would not be selecting *research* at all; we would be selecting *people*.

The operation of a people-centered approach turned out to be remarkably easy. To be honest, we really had no alternative. This is simply because the process of judging *proposals* inevitably traps the judges into the classifications of fields or disciplines. When funds are scarce, the judges cannot avoid becoming entangled in the jungle of bureaucracy and red tape outlined in Chapter 5. We treated selection as if it were a scientific problem rather than an administrative one and struggled constantly to reduce the number of rules we imposed on our applicants. One must think and work as a scientist to operate such a scheme. In contrast to other scientists, though, our work would be entirely on the conceptual plane. Since we covered the entire spectrum of science and technology, we could not descend into the undergrowth of detail because we would immediately become bogged down. We also had to be willing to forgo the satisfaction of personal discovery and to take vicarious pleasure in the insights of others. We had to have applicants' full cooperation, of course, which meant that they must want to take part in our novel scheme and be willing to endure the new procedures. Recalling that we were striving for objectivity, our problem then became one of how to remove the subjective elements as far as possible and create an environment in which *pioneering scientists could select themselves.*

Our remit from BP was to support the research that might lead to new types of industrial activity. The word "new," however, is not particularly specific. Almost every industrial company in the world is looking for new opportunities, and they collectively spend many tens of billions of dollars a year on their searches. Our budget would of course be only the minutest fraction of the global investment in research, so how could we hope to make a difference? Our attempts to answer that question led us to develop the concept of *the relative value of money.* If our few million were earmarked for research in, say, AIDS, cancer, or superconductivity, it would indeed make little difference to the billions being committed annually to such mainstream fields. On the other hand, if our modest funds could be used to tackle important problems that nevertheless were being virtually ignored, then they could become much more valuable. But the concept of relative value had another implication that took us a year or two to recognize. Venture Research would only tackle problems of substantial importance. Success would mean that the rewards could be so high that the initial outlay becomes almost irrelevant. Viewed in this light, the question of the cost of research becomes an artificial constraint, and we were trying to get rid of those. This meant that if we were serious about deriving as objective a strategy as possible, we should take the astonishing step of banishing financial issues from our selection criteria! In effect, we should adopt what seemed to be the outrageous policy of acting *as if we had an infinite amount of money at our disposal.* See Text Box 34.

It should not be too surprising that the penny took so long to drop. Our unit was immersed within an industrial company and cost is a constant preoccupation. No limit had initially been set on the funds we might spend, but my view was that we should not press too hard on that front. Many in BP thought that what we were doing was altruism, "an elegant way of spending a modest amount of money," as one executive put it. But they would be happy only if our expenditure remained modest. (Alas, we were not able to convince some of them that our objectives were much

> ## Text Box 34: A Venture Research Story
>
> One of our most successful research programs in financial terms was Steve Davies' at the University of Oxford. His first grant in 1985 was for some £150,000 over three years. The second for the subsequent three years was for some £270,000. His ideas were commercialized through a new company—Oxford Asymmetry—floated in 1998 on the London Stock Exchange at an initial valuation of £200 million. A great deal of further investment together with considerable commercial expertise was required, of course. However, the full cost over the six years of the initial university research on which the development was founded was some 0.2% of the company's valuation at flotation, thereby providing post facto justification for our policy of not allowing research costs to influence our decisions. Furthermore, ours was not a scattergun approach. The people we backed were almost certain to succeed scientifically, and most of them did (see Appendix 1). The remaining question was whether new science leads to new industry, and that risk too can be managed.

wider than benevolence alone. It was not until well into the 1990s that our operations could be seen in retrospect as having been highly profitable, as we knew they would be eventually, but by that time we had ceased to exist.) Had we presumed at the outset that we actually had access to unlimited or even substantial resources, it would have been seen as profligacy, and our days would have been numbered. On the other hand, had we cut our cloth according to the money we thought we might have, at any level of funding, we would never have had enough. That policy leads inevitably to the very traps described in Chapter 5. The advantage of our implicit assumption of unbounded wealth was that finance would be removed from the equation, and we would be forced to concentrate exclusively on scientific issues. That was the big difficulty, of course, and we knew we would be making rods for our own backs because stringent scientific criteria are much more difficult to satisfy than financial ones. Indeed, our apparently profligate policy turned out to have the advantage that we never fully spent our projected budget allocation in any of the 10 years of the unit's operations! Really good ideas are really hard to find.

The details of our selection criteria slowly began to crystallize out. The unit had been given great freedom. We would therefore search for the scientists who needed freedom from bureaucratic constraints more

than anything else, that is, freedom to launch radical challenges to conventional thinking in any field or start something entirely new. Final selection would be by face-to-face-dialogue, which is a much more direct and reliable way of assessing potential than reading proposals. Our most immediate task, however, was to generate proposals, and that is not as easy as it may sound. We could not advertise in the media because we could not then control the number of applicants we would have to deal with. The unit, shown in Figure 15, was small and could easily have been

Figure 15
Venture Research Unit in 1990. Clockwise from left: Jean Shennan, Tony Regnier, Julie Fleming, David Ray, Sally Champion, Fionna Foy, Tim Sanderson, and the author. We are about to cut the cake baked for our tenth anniversary by Jean Shennan. (Photograph by the author)

swamped.* We took the exhausting alternative, therefore, of promulgating our mission on the hoof. We visited almost every university in Britain and a large number in mainland Europe and North America, gave talks, and simply requested proposals.

It was not merely a matter of selecting a university and turning up. Each visit took weeks to arrange. One of our problems was that there seems to be little common ground between the industrial and academic sectors. The climate was such that universities were intensely suspicious of our motives. Our correspondence was often misunderstood. Some universities seemed to suspect that we were trying to steal their best ideas from under their noses. Our enquiries came on BP-headed notepaper, of course, a fact that made it very difficult to convince them that the unit's interests were at least as wide as any universities. Apparently, they would read only the letterhead! Some universities would insist on selecting the people we could speak to. This often meant that we had to listen to what the university thought were new ideas for getting oil out of the ground! It often took several visits before it became accepted that we were not interested in oil at all, and we finally got to speak to those we might be able to help. The key question we asked was: If you had all the money in the world and the freedom to match, what would you do that you are not doing now? We knew, of course, that only a tiny number of applicants would qualify for our funding, and so we placed as light a burden on them as possible. Proposals should be less than one page or could be discussed with us there and then. There were no other rules and *anyone* could apply.[†]

After about two years, we were receiving one or two formal proposals every working day—about 300 a year on average. At each university visit, however, we would usually get up to 50 enquiries, each of which could have been converted into a full-blooded proposal had we offered encouragement. Scientists had nothing to lose by contacting us. They did not even have to tell their universities—heads of departments, for

*The unit grew slowly throughout the decade, and had reached eight members, including all secretarial and administrative support, at its close in 1990.
[†]As I have mentioned earlier, the conventional academic-funding organizations will rarely accept research proposals from scientists without tenure, and postdoctoral researchers are generally excluded from making direct application for independent research support. Nevertheless, the VRU received few proposals from postdocs even though we went out of our way to stimulate them. One postdoc whose proposal we had agreed to support later withdrew. His faculty had warned him that the new territories our substantial grant would enable him to explore could put his career in jeopardy!

instance—that they were doing so. That step would only be necessary if the scientists had been successful in winning our support. This turned out to be a wise move because scientists could privately share their innermost thoughts with us without having to go public, so to speak. Encouraged to push their ideas to the limit, they sometimes even surprised themselves. We always tried to give an immediate response to applicants, especially to those we met during our visits, but we sometimes fell short of that. As might be expected, our most frequent response would be that the ideas as outlined seemed to be extensions of what the scientists were doing already. Such next steps could be funded by the conventional authorities and therefore did not need the unlimited freedom we could offer. Our offer to provide unlimited resources to those who needed them led to many opportunistic bids for very expensive equipment or the provision of small armies of researchers. See Text Box 35. Such requests, we would simply explain, imply that you know precisely where you want to go and what you want to do, and that is not consistent with an exploratory venture into the unknown. We would always add, however, that if we had misunderstood, to please let us know.

Almost every proposal we received involved perfectly respectable science, but that was not all we were looking for, of course. There had to be some hint of heresy, some glint of the unexpected, before we would start getting interested. As soon as we could arrange, therefore, we would

---

Text Box 35: Soaring Ambition

Perhaps the most audacious and imaginative proposal we received came from Alvaro de Rujula, a theoretical physicist from the international laboratory for nuclear research—CERN—in Geneva. He wanted funds to build two enormous proton accelerators (at least 1000 GeV) in Earth orbit, arranged so that each orbit was perpendicular to the other. The protons would be used to generate large fluxes of high-energy neutrinos, thereby bathing the whole Earth with them. Detectors would be placed at strategic places on and below the ground all around Earth. A positive signal would indicate the presence of hydrocarbons as the probability of a neutrino hitting a proton is higher than that for any other particle and hydrocarbons are prolific sources of protons! His experiment should therefore reveal the whereabouts of *every* gas or oil field on the planet. The cost would be many, many billions of dollars, of course, but that, he said, would only be a year or so of global oil and gas revenues!

let the applicants know what we thought, but we did not close the door on anyone. They could and often did come back if they thought we had missed the point or they had forgotten to explain something. Lifelines like these can be crucial. It is very sad that most (if not all) conventional agencies will not allow second bites at the cherry. Rejections from these bodies are usually terminal, and proposals may only be reconsidered if they have undergone radical revision and then only after some extended cooling-off period. We had no such inhibitions. Whenever we heard about a proposal that seemed unusual (which were about 10% of the total), we invited British scientists to come to our offices in London, at our expense, so that we could hear first hand what they had in mind. Those who had sent proposals from overseas were usually visited by one of us during our next trip.

We are now coming to the most distinctive feature or our selection procedures. Funding agencies rarely use face-to-face selection. There is no good reason for this. Bureaucratic procedures thrive on the written word because it is much easier to classify, grade, and file. Decisions too can be transmitted in formal standardized prose so that one does not have to actually meet the person whose hopes you may have just destroyed and explain why. But it is very difficult to express one's thoughts clearly in writing, and scientists are not especially good at it. Why not allow them to explain their ideas in their own words? It is something they must do every day, and most academics excel at it. We decided to let applicants play to their strengths, therefore.

Sadly, however, one of our most difficult problems at our meetings was to persuade prospective Venture Researchers to concentrate *exclusively* on their science. Usually, they came prepared to discuss a wide range of extraneous or irrelevant issues and needed frequent reminders to return to basics. Thus, they would normally begin by outlining the possible goodies their research might create, objectives that might be fine for mainstream research but would be unlikely to lead to the unimaginable breakthroughs we were looking for. The usual enquiries about how much money we might provide could be dealt with quickly, but it was more difficult to discourage them from trying to say the things they thought we wanted to hear rather than telling us what they really wanted to do. These problems often meant that it took more than one meeting before we could begin to make progress—that is, before we could bring them to consider trusting us.

It is astonishing that so few industrialists and other investors in research seem to understand the importance of trust. In my experience, the consensus among investors seems to be that trust is irrelevant if one takes the best legal advice and arranges that the proposed research is protected by bullet-proof contracts. Such investors would be better advised to leave their money in a bank. When scientists or engineers are bound by rigid contracts, they will do precisely what they are contracted to do—neither more nor less. If that is what the investor wants, then fine. But where is the research in that? If by chance a researcher comes unexpectedly across the crown jewels, so to speak, it is virtually certain that a rigid contract will not have anticipated them. In those circumstances, the researcher will probably, and justifiably, take his or her wondrous discovery to the highest bidder. The investor will get only what has been paid for.

Trust, like love, is intensely personal, of course. Thus, it is scarcely possible for a committee to trust another committee or for a committee to trust an individual. The terms of agreements when committees are involved must therefore be precisely described—hence the paraphernalia of controls described in Chapter 5. But research is an intensely personal activity. Facts have little meaning unless they can be interpreted and placed within a framework of comprehension. Insights, flashes of genius, and eureka moments in general are accorded only to individuals. Researchers may work very productively in groups, but individuals initiate discoveries. Individuals can inspire others. Groups whose members trust each other can work miracles. But trust cannot be fostered to order. No amount of money can command it. Mutual respect seems to be the first step, but however it is done, trust must be earned.

Mutual trust was therefore our primary goal. Without trust, our visitors would probably not reveal what they really thought about the state of their particular science and what should be done about it. We encouraged them to tell us about the significance of what they were proposing to do, to put their work into a global context, to explain the concepts behind their ideas, and if they thought that the levels of understanding in their science were weak or seriously flawed, to explain what the consequences might be. Seated around our little round table, if we could agree that their work could radically change perceptions on something very important, they were the people for us. The arrival of that rare moment was always magical and never needed debate. It became obvious spontaneously, like the turning on of a light, although it may have taken many hours to find the switch. From that moment, we in the VRU joined the scientists. We were in collaboration and would do everything in our power to help bring the scientists' vision to fruition.

We also had to build up our trust in the scientists. If we were to support their research, we would have to trust them not to fritter our valuable "free" funds (see next paragraph) away on things they could have done with conventional support. This point was particularly important because their only real obligation would be to keep us informed about what they were doing and even that requirement was, strictly speaking, not enforceable. There had to be legal contracts, of course, but BP's legal people were clever enough not to tell us what to do. This was a major stroke of luck because all too often large organizations insist on standard contracts being used. Such heavyweight documents rarely allow the freedom and flexibility essential to research. For genuinely open-ended research, which is of course what all academic research should be, there can be no specific objectives against which to gauge progress. There are no milestones in the wilderness. Trust played a very special role, therefore. Once the researchers had our money, it was an essential part of our strategy that they should be free to use it as efficiently as possible. Circumstances change, and they should be free to respond to them. Therefore, we had to convince ourselves that we could trust them to do their best and tell us about it. We had to persuade them that they could always trust us to back them to the hilt whatever happened.

I mentioned that our funds were "free." Readers outside the funding business may not understand this remark, but it should not need to be made. Let us say that you have won an award to carry out research whose costs have been carefully calculated and agreed upon with your sponsor. That is, you have been through the agonies of creating a viable research proposal, have prepared it so that it satisfies all the relevant bureaucratic rules, and have piloted it through the perilous pathways of peer review. These tasks alone are formidable, and one might think that you should now be left to get on with the job in hand. But that would be far from the case if you have won the support of a conventional funding agency. Funds are usually "earmarked"; that is, they are tied to specific expenditures on such individual items as equipment or salaries and made according to strict timetables. Both the universities and the agencies have arrangements for policing these strict rules. Variations are sometimes allowed, but they must be agreed upon in writing. *Any funds unspent within the time specified are normally lost.* It is, of course, a crazy way to treat responsible scientists, and it encourages waste. One's departmental standing can depend on the accumulated total of one's grants, but the funds only count toward one's total *if they have been spent.* Approved funds for which the need might have passed, say, or which might spill over the agreed-upon period, are therefore spent on *anything* that might qualify, whether or not it is needed.

Free funds can be spent as if they had been deposited in the scientist's bank at the beginning of the award period. They can be spent at any time. Accountability is not lost by this simple expedient as invoices must still be provided. We usually asked to be told about major changes to the initial proposals, but a telephone call would do. This freedom allows the researcher to respond to events—events that of course will arise unpredictably in a genuinely exploratory initiative—and enhances the relative value of the funds. From the sponsor's point of view, value for money increases accordingly. It is difficult to quantify the advantage as comparisons cannot be made. However, one might expect the earmarked regime to work best when the research runs precisely to plan. That should rarely be the case for exploratory research. Several of our Venture Researchers claimed that the simple expedient of making our funds free roughly doubled what they could do with them. Conventional agencies could therefore substantially increase what can be done with their present budgets merely by moving toward a free-funds regime.

To return to our main theme: After the magic came the hard work. People who change perceptions on important problems eventually change what is done. They might therefore lead to the creation of the new industry we were searching for. The next step would be to discuss the best means of achieving their scientific ends and to estimate the resources needed for the fastest progress. In our experience, armies of assistants were always unnecessary. Venture Research is subject to considerable uncertainties, and time to think is usually more important than pairs of hands. Similarly, very expensive equipment was usually unnecessary because the scientists were going into virgin terrain, which from a scientist's point of view is the ultimate target-rich environment. Determined scientists can hardly avoid making discoveries in these circumstances, and the simplest equipment is usually adequate. One hears a great deal nowadays about lead times and milestones for even so-called basic research. For Venture Research, however, these concepts are meaningless. No specific objectives are set. Once one enters virgin terrain, discoveries are likely at any time, including the earliest phases of the exploration. One cannot predict *what* they will be, of course, or whether they will be valuable.

The VRU had no money of its own, and so once we had made up our minds, we would collaborate with the scientists on a description of the proposed work. This was attached to our recommendation to the VRAC. Proposals were usually short as they concentrated on the concepts involved and were light on detail because the means of achieving the broad objectives would almost certainly change as work progressed. The objectives themselves might also change. It had to be a joint effort because scientists have little experience of producing this type of proposal. It was not uncommon to go through 10 drafts before we were eventually satisfied! We had to be sure that the final proposal summarized their ideas as succinctly as possible because the VRAC never met the applicants.

One of our biggest difficulties arose from the VRAC's insistence on an "independent" assessment of the proposed research. They had never been impressed by my criticisms of peer review, and I soon stopped trying to persuade them to forego its doubtful value. The compromise we reached was that we would include as part of the final proposal a written opinion from one of the most distinguished workers in the general area of the proposed research. I usually selected the VIP in consultation with the researchers as being the person most likely to give his or her support. I would then speak to the person, usually by telephone. I would explain what the unit was trying to do, summarize the research, and try to coax the person into endorsing it in the most enthusiastic terms. My request would be confirmed in writing. Their written opinion came to me first, and if it seemed that they had got hold of the wrong end of the stick, we would speak again, and I would try to persuade them to change their minds. After such comments as "this is not the way things are usually done, you know," they often did see our point. But they did not always do so, and we had to allow them that freedom. As we had had the benefit of many hours of discussion with the researchers, our minds were made up at that stage, so we were looking only for the "peer's" endorsement. Their remarks were often helpful and perceptive and sometimes led to our making changes in the proposal. This is peer review working at its very best.

Occasionally, however, our use of peer review was an arrangement bordering on the farcical. We sometimes recommended a proposal to the VRAC to which was appended severe criticism from the peer along such lines as the idea would not work or the details were inadequate or the scientist was no good. One carefully selected peer's assessment of what turned out to be one of our most successful scientists (Ken Seddon, see below) contained all these objections! Most importantly, however, I was a member of the VRAC, and I could argue the scientist's case. The VRAC, to its great credit, almost always backed the unit in these circumstances.

The VRAC met some three times a year, usually for about three hours. It rarely had more than a handful of proposals to consider, so each one received rigorous attention. In the weeks before a meeting, I would spend a few hours with each council member to brief him on the research we would be recommending and to get his agreement on the selection of the peer. Once the peer was selected, we would have no alternative, of course, but to convey his final opinion. These premeeting meetings also provided invaluable feedback, and we could therefore add a little more fine tuning to the proposals in consultation with the scientists.

I met the VRAC chairman, Jim Menter, more often—usually about once a month. One of our meetings is shown in Figure 16. We had an

*As I explained in *To Be a Scientist* (Braben, 1994), this process as applied to research proposals should strictly speaking be called *peer preview*.

Figure 16
Sir James Menter and the author at the Venture Research Unit's office in 1989.
(Photograph by BP. Reproduced by permission.)

unusual collaboration, the laconic and the loquacious as we were. He was a man of very few words and never used one syllable when none would do. I had to learn to read the other signs as we hammered out our strategy, and one of the first was that his silence, as he puffed on his small cigar, did not necessarily mean consent. We rarely disagreed. That is, he never said he disagreed, although lesser men would have done so. His main concern seemed to be whether what I wanted to do would be effective in achieving our objectives. There is no point being right if no one will listen.* I often found our meetings infuriating because of the lack of verbal response, but I always found on reflection that his apparent lack

*During my travels in the United States, I came across the following verse which illustrates his point very well:

> Here lies the body of William Jay
> Who died maintaining his right of way
> He was right, dead right, as he sped along
> Now he's just as dead as if he were wrong

of enthusiasm would mean that I had not thought my case through, and so I had better sort things out before our next meeting. In this respect, therefore, he was for me the perfect foil.

The derivation of these apparently simple rules and procedures for evaluating researchers and their ideas took a long time. In retrospect, I can now see that our biggest problem was that we were starting from scratch. There was no experience on which we could draw. Academics in the old days generally did what they pleased. They had to get themselves appointed, of course, but that was a job selection process much like those in other professions. Once appointed, their research was not usually subject to second guessing. With hindsight, our problem was to recognize that virtually every selection protocol used nowadays inhibits adventurous research involving major departures from the beaten tracks. But that was precisely the type of research we were looking for. We had therefore to go through a process of unlearning, of weaning our minds from dependence on methods of selection that seem to have been written on tablets of stone. Not surprisingly, it took perhaps four or five years before we became confident that we really knew what we were doing and the new procedures could be operated smoothly.* Passage through the VRAC, however, was never smooth, and we were severely punished if we had not done our homework. The collective disapproval of such a powerful group was an agonizing experience that could not be endured too often, which, of course, was a strong inducement to get it right the first time. Happily for us, the VRAC turned down very few of the unit's recommendations in what turned out to be our last five years. I was not convinced that they were right on those rare occasions, but rough justice or not, we had to accept it.

Praise for corporate BP in providing the environment for all this cannot be overstated. BP was my employer, of course, but my anomalous position did not fit readily within any corporate structure. BP was more accurately my patron, and provided I kept to the rules I had been instrumental in creating, any reasonable request for freedom or resources was usually approved. The unit was never exempted from the many and frequent assessments to which every BP department was subject, but there was virtually unanimous and unvarying corporate agreement on our ob-

---

*The unit's expenditure during its first five years was much less than half the total for the decade. The research supported during that first period was certainly of the highest quality, as otherwise the VRAC would not have authorized it. Its coefficient of adventurousness, so to speak, was generally much less than that for our second five years because by then we knew precisely what we were looking for.

jectives. We had only to demonstrate that we were moving toward them as efficiently as possible. It helped that a wide range of the most senior BP staff took a close interest in what we were trying to do, mainly because, I think, it appealed to their imaginations. I worked hard on that,* of course, and often invited them to our conferences in Britannic House, BP's London headquarters, even though few of them had scientific backgrounds. They rarely came along to the meetings but sometimes joined us for our buffet lunches or dinner. It was a dangerous game, because we had to allow them to mingle among our researchers unchaperoned. As one BP VIP half-jokingly remarked, "Yes I'll come to your gathering. I want to see the sort of people you're wasting BP's money on." Nonscientists or not, they could all tell a hawk from a handsaw, and we would not have lasted long had they not been impressed.

In 1982, Jack Birks retired, as did Robert Belgrave, and so I lost my original patrons. This can be a dangerous time for adventurous protégés, but I need not have worried. Roger Bexon succeeded Birks briefly, but after a few months, BP recruited Robert Malpas (later, Sir Robert) from outside the company to take over Birks's responsibilities. He was BP's first managing director who had not been home grown.

It was an inspired choice. For nearly seven years Bob Malpas, who is pictured in Figure 17, bore BP's entire technological mantle, under which enormous cloak the miniscule Venture Research Unit also took shelter. In spite of the unit's size, Malpas continued with the arrangement set up by Birks by which I reported directly to him,† as did BP's director of research and its chief engineer. Thus, about once a month we had a one-to-one meeting to discuss the research that we were doing or might be in prospect, and generally to review our progress. Needless to say, Malpas was a powerful personality. He was also hyperactive. One never seemed to have his undivided attention for more than a few minutes when he would suddenly leap up from the conference table to attend to some pressing matter that had just occurred to him, returning as if we had not left off. He had an extraordinary grasp of matters technological and had been renowned at ICI as well as BP for his attention to "the plant after next," which he considered essential to a company's long-term survival. Our

---

*Our *Entertaining Science* series of lunchtime lectures held in Britannic House, which were open to anyone, usually attracted audiences of around 200. The topics presented to the general staff included quantum gravity and the distributed intelligence found in ant colonies.
†In BP, as with other large multinationals, a managing director would not normally be directly involved in activities where the expenditure was less than about $100 million pa.

Figure 17
Bob Malpas, the Venture Research Unit's contact managing director, is on the right. He is shown with Jim Menter, center, and Hugh Norton, a BP managing director who also had close links to the VRU, at the party to mark Jim Menter's retirement from BP in 1989. (Photograph by BP. Reproduced by permission.)

discussions on science and technology were as interactive and invigorating as any I have had. When it came to policies, however, we sometimes had robust disagreements, and I feared for my job on more than one occasion. But when our time was up, he would usually conclude with a broad smile of encouragement, an arm on my shoulder, and the question of when we would next meet. These occasions were stimulating for me, too, and ensured that the unit was kept on its toes.

As the VRU's program flourished, it became increasingly clear that BP could only ever be interested in a small fraction of our output as we covered the entire spectrum of scientific fields. The idea that Venture Research should be financed by an industrial consortium led by BP and made up of noncompeting, compatible companies with complementary interests seemed the most obvious solution. It would reduce BP's contribution by a factor of up to about 4 or 5—according to the number of companies involved—without affecting BP's a la carte choice from our menu. Electronics (possibly Sony) and pharmaceutical companies (possi-

Figure 18
Sir Hans Kornberg, Master of Christ's College Cambridge, right with Tim
Sanderson from the Venture Research Unit, left, and the author, at Jim Menter's
retirement party in 1989. (Photograph by BP. Reproduced by permission.)

bly ICI or DuPont) seemed to offer the best opportunities for a partner-
ship. BP Finance, a newly formed BP company, had almost completed
the structural arrangements for this wonderful initiative when we were cut
short by the cruel corporate axe.

Malpas had finally been persuaded late in 1988, just before he was
to retire from BP, to relinquish his direct responsibility for Venture Research. He took this decision despite the most passionate protests I could
dare to make. Malpas was clearly concerned, however, and took the unusual step of coming to a VRAC meeting to report on the new arrangements, a gathering that Sir Hans Kornberg, pictured in Figure 18, described with his usual urbane wit as a Royal Flush. Malpas told the
VRAC that managing directors had just concluded a review of the shape
of research within the company. Cadogan would now be chief executive
of the newly formed BP Research International, and also of BP Ventures,
BP's venture capital arm. They had also looked at the VRU. Managing
directors (MDs) had concluded, he said, that Venture Research was an

excellent idea. It had been done remarkably well and was beginning to bear fruit. MDs wanted to harness that success, which they had concluded would best be done if Venture Research were to be included within BP Ventures, which in the future would also be responsible for its funding. There was no intention, he went on, of placing Venture Research "in a straitjacket," but the new arrangements would have the advantage of easing the interface with development.*

Jim Menter commented that it seemed a logical arrangement, and so it was. He had told me privately that although he knew I did not like it, he was sure I would do my best to make it work. I should have been ecstatic. Malpas had just delivered an accolade that I could hardly have dreamt of eight years earlier. Yet, despite my respect for Malpas and his fellow directors, I was filled with dread, and I wrote in my occasional diary that we had probably come to the end of the line. I knew from bitter experience that Cadogan did not seem able to tolerate individuality, and that trait above all is an essential ingredient for Venture Research. I believe he also saw Venture Research more as a threat than as an opportunity. However, I was now to be one of "his lads." At my first meeting with my new colleagues, my enduring memory is of Cadogan breezily turning to Howard Lunn, his deputy at BP Ventures, to ask him, "what problems do you want me to solve today."

The end came from a most unfavorable conjunction of events. Menter retired from BP in late 1989. He was succeeded on BP's main board by Sir Robin Nicholson, whom I knew well from his time as chief scientist in Whitehall. He was also to take over at the VRAC, but as things turned out, he never did. Sir Peter Walters, who had been chairman of BP since 1981, announced that he would retire on April 1, 1990, but handed over the reins to his successor Robert Horton (later, Sir Robert) some weeks before that. Needless to say, Walters was an amazing person. He had run one of the world's largest companies very successfully for nine years yet he had an astonishing lightness of touch. Always relaxed and unassuming, he had the gift of being able to address the largest gatherings as if he was having a personal conversation. Nevertheless, he was formidable. One of his senior colleagues said that at the regular meetings of

---

*This essential condition had never been satisfied in the past. The unit had already spawned a discovery of a new type of medical diagnostic kit—amplified enzyme-linked immunoassay (AELIA)—but BP had no interests in the medical field, of course. The discovery was developed elsewhere, and when it was sold in 1986, its value exceeded BP's total integrated investment in Venture Research up to that date. Regrettably, the condition was not to be satisfied in the future either.

MDs, he had them all "sitting on the edges of their seats." We were lucky that he had always looked favorably on the unit,* and indeed Malpas' accolade must have had his full endorsement.

As 1990 dawned, I was beginning to think that we might have a future after all. The Sony Corporation in Japan had expressed strong interest in joining our proposed consortium. BP had appointed Saatchi and Saatchi to develop an advertising campaign under the banner *For All Our Tomorrows.* Saatchi and Saatchi had proposed that Venture Research should play a prominent part as they thought that Venture Research encapsulated BP's vision of the future, with its emphasis on people and new ideas. Our research program was blossoming, and we had arranged a tenth anniversary conference for the summer that was to be attended by around 250 of the scientific great-and-the-good from Europe and North America. The Sony Corporation was also sending representatives from Tokyo, as were the British national press.

I could not have been more wrong. BP's change in chairmanship heralded a new corporate culture. Core business was now to be given *absolute* priority, and Venture Research was apparently incompatible with BP's new philosophy. Following a meeting of MDs, I was told on March 7 by a jet-lagged Basil Butler (Malpas' successor) phoning from New Zealand that Venture Research had to go.

The Venture Research conference held on June 26–27, 1990, was the most successful we had held. We got almost half-a-page in *The Financial Times* and good coverage in *The Telegraph* and *New Scientist.* Our three Sony guests were clearly impressed by the quality of our Venture researchers. As soon as the conference ended, however, I was caught up in a whirlwind of frenetic activity. It involved much legal hocus-pocus, and the timetable for my departure seemed to be set in stone. However, BP agreed to honor all its contractual obligations to our community of Venture Researchers over the coming three years or so, but I would not be involved. Thus, a noble, glorious, imaginative, and far-sighted experiment came to an abrupt end. I left BP on July 31, 1990, with a small pension and a generous financial contribution toward the legal costs of setting up Venture Research International (VRi) and the new Fund.

On August 14, 1990, President Ohga of the Sony Corporation in

---

*Peter Walters once attended one of our annual Venture Research conferences at Britannic House, spending four hours with us when we had expected him to stay for as many minutes. These conferences were remarkable in themselves. Held in London, they were exclusively dedicated to the most profound aspects of science and engineering and covered almost every discipline.

Tokyo* confirmed its commitment to Venture Research if we could find two other companies that would match it. On September 7, the United Kingdom's national channel 4 television dedicated an entire one-hour *Equinox* program[†]—*Blue Skies*—to the story of Venture Research. The estimated audience was 1.4 million. Sir John Fairclough, the retiring chief scientist in Whitehall, had accepted the chairmanship of our new VRi board.[‡] This rosy picture soon began to fade, however. In late 1990, Ron Coleman, Chief Engineer and Scientist at the Department of Trade and Industry, had offered to match from public sources any funds we could attract from two or more U.K. companies, thereby reducing its subscription to our consortium by half. Unfortunately, we could not find even one. I visited a wide range of companies in Europe, Japan, and North America over the following few years, but perhaps because of a deepening economic recession, there was lots of polite interest but no further commitment.

The decade of the 1990s continued to be frustrating, but professional optimists are not easily daunted. John Fairclough suggested that our fund-raising prospects would be enhanced if VRi included an executive with commercial experience, a Rolls to my Royce, as he flatteringly put it. In February 1995, Iain Steel, who had been chief executive of BP Ventures during the 1980s, was the ideal candidate, and he accepted an invitation to join us even though, as for other board members, we could offer no immediate financial compensation. He would do it for the fun, he said. His experience was just what we needed, as he had played a key role in a number of substantial and successful commercial developments arising from BP's mainstream research. Just as important for me, however, were his wonderful sense of humor and down-to-earth pragmatism that have often proved invaluable in getting me through the darkest days.

Despite the growing evidence of the scientific and commercial success of our unusual methodology, we have been unable to raise the funds

---

*The Sony philosophy on new associations was to enter gradually at first but to expect the scale of its commitment to increase. Thus, it decided (on August 14, 1990) that the appropriate level would be a minimum of $1 million pa for three years, which it was prepared to review as the demand increased.

[†]Made by World Wide Films, London, 1990.

[‡]Other members at that time were Nigel Keen, Chief Executive of Cygnus Venture Partners; Dr. John Hendry, Director of the MBA Course, Judge Institute of Management Studies, University of Cambridge; and Dr. David Ray, who had been the Deputy Head of the VRU. Professor Sir Harry Kroto of the University of Sussex designed the Venture Research International logo. He received the Nobel Prize for Chemistry in 1996.

needed to re-create the glorious experiment that ended in 1990. Common sense might say that we should give up after devoting so many years to the crusade. However, the intellectual prize is so great that quitting seems unthinkable. The initial cost of the initiative would be modest. BP's total investment in Venture Research for the entire decade of the 1980s was some £15 million. So far its actual monetary value would seem to have increased more than 20-fold. Its full scientific potential remains mainly untapped to the present day. With just a small dash of luck, we might find that visionary patron who wants to buck the trends and do something really original and exciting.

Some of the scientific and other successes from the 26 groups of researchers we were supporting in 1990 are briefly summarized in Appendix 1. Most, if not all, the 26 achieved substantial scientific success, and many achieved what might be called "scientific breakthroughs." These words are often overworked, and many scientists might regard them with suspicion as being unduly emotive, especially when applied to their own work. Breakthrough here is used as a shorthand to describe work that has succeeded in radically changing perceptions, drawn widespread attention to areas not previously regarded as important, or met scientific goals that were generally thought to be impossible.

The new initiative would use exactly the same criteria for research selection as we used for the old Venture Research. Fund raising is never easy, but one might think that with such an impressive track record it should not be overwhelmingly difficult. One would be wrong. We came agonizingly close on four occasions since 1990, but each bid eventually failed at the very last hurdle. There seemed to be two main reasons, but one never knows, of course. Reasons are never given. Discrete enquiries revealed, however, that investors were uncomfortable with the lack of control. I do not mean financial control. That obviously would be given high priority. However, would-be investors wanted to know from the outset what research we would support and what the outcomes might be. How times have changed since Thomas Watson backed his "wild ducks"! (See Text Box 22.) We would offer, as did Watson, freedom to explore the unknown—not to anybody, but to inspired people with an outstanding idea and a flair for successful exploration. No one can know, therefore, not even the researcher, what completely new things might be uncovered.

A second reason for investors' lack of enthusiasm seems to be based on the assumption that there is an abundance of schemes purporting to support "Blue Skies Research." It might be presumed that ours is not new, therefore. Unfortunately, the term Blue Skies Research is not only vague but has become meaningless. So, we stopped using it for a time and used the term Breakthrough Research instead. But these words too have become overworked. They also sound somewhat pretentious, so we returned to the original Venture Research, by which name we are best known. This also has its disadvantages, as nowadays, the word *venture* immediately spawns the word *capital*, again giving the completely wrong impression of what we are about.

One or two words will never be enough, of course. But we love the Blue Skies sobriquet and are doing what we can to give it meaning while retaining its magical attraction. For us, Blue Skies Research is useful shorthand for the most innovative and creative research that might lead to outcomes unimagined. This means that the research should have no specific or directed objectives. The "skies" should initially be clear and full of promise. Any imagined outcome should be possible. However, these sentiments run counter to current practice in the Blue Skies Research game, due, of course, to the dead hand of peer review. Since the conventional agencies cannot normally release funds unless they have the endorsement of this sacred cow, disciplinary constraints must be imposed. How else can peer review work? But in our view, to speak of Blue Skies Research in the field of, say, cancer or any specified objective, as is usually the case, is a flat contradiction. Such initiatives merely play lip service to the idea of fostering the full range of human creativity.

As should be much better known than it is, quality cannot be quantified. Those who work in the bureaucratic jungle do not agree, apparently, as we have discussed. There are also national schemes claiming to support Blue Skies Research that do not even come close to deserving that lovely accolade, although they may sponsor excellent research. Venture Research, that is, the route to our version of Blue Skies Research, is exclusively dedicated to uncovering the unknown and seeding the blue skies. It is vital, therefore, that we differentiate what we are trying to do from the multitude of mission-oriented endeavors now extant. Following current jungle fashion, therefore, it may be instructive to propose a novel way of measuring the skies' blueness in a research proposal. The Braben Venture Research Index—BRAVERI*—offered here is intended to be a rough rule of thumb. *It is, of course, not to be taken too seriously.* The Index is based on ten features that might be used to describe a research proposal. They are not independent. Each feature yields a score of 0, 1, or

---

*The index was presented at the 60th birthday meeting in Messina, Sicily, to honor Gene Stanley, one of our American Venture Researchers. An earlier version was published in *Physica A 314*, 768 (2002), and is reproduced here by permission of Elsevier.

2. The BRAVERI level is found by adding together the scores in each category, and expressing the total as a percentage—that is the purest Blue Skies Research would score 20/20, or a BRAVERI of 100%. The features are:

1. The proposed research:
- Is within the purview of a national funding committee, priority, or initiative. If so, score 0.
- Straddles two national funding committees, priorities, or initiatives. If so, score 1.
- Extends over several national funding committees, priorities, or initiatives; or none are relevant. If so, score 2.

2. The objectives of the research:
- Are detailed and specific. If so, score 0.
- Are general. If so, score 1.
- Have not been externally imposed. The researchers are completely free to tackle difficult problems in a general area in any way they think fit. If so, score 2.

3. Assessment of a proposal by peer review:
- The reviewers are fellow experts. Their opinions are pertinent. If so, score 0.
- Involves several experts from different fields. It may be difficult to find a consensus. If so, score 1.
- Is problematic in principle, because the researchers are radically challenging accepted thinking or proposing something totally new. Genuine peers may be difficult to find. If so, score 2.

4. Funding requirements:
- The probability of success would increase as the funding level increases. If so, score 0.
- The probability of success is not sensitive to funding level. If so, score 1.
- Success is not easy to define. Expensive equipment may not be required. Thinking time is essential. Only one or two research assistants can be efficiently employed. If so, score 2.

5. Implications of the expected results:
- The most likely outcome is a useful addition to knowledge. If so, score 0.
- They could lead to a breakthrough in the development of a field. If so, score 1.
- They could be revolutionary, and radically change accepted thinking. They could lead to the development of new fields. If so, score 2.

6. Timescale:
- The research is expected to meet its targets in the time allowed. If so, score 0.
- The research is part of a long-term program. The ultimate goal is unlikely to be achieved in the time allowed for a specific phase. If so, score 1.

- Is indeterminate. Successes, whatever they are, could be achieved at any time, or never. If so, score 2.

7. Competitors:
- The researchers are striving to be first to a specific goal. The competition is fierce. Cards should be played close to chests. If so, score 0.
- The researchers are striving to be first, but the field is wide open and direct competition is unlikely. If so, score 1.
- The researchers are striving to understand. The research is probably unique. There is therefore little or no competition. If so, score 2.

8. Publication:
- The expected results will probably be published by a mainstream journal. If so, score 0.
- The expected results might be published by a prestigious journal such as *Nature* or *Science*. If so, score 1.
- It is not clear at the outset what the results might be. When they come, it might initially be difficult to get them published. If so, score 2.

9. Prizes:
- The researchers do not really expect to win a prize. If so, score 0.
- It is conceivable that the researchers might win an award from a learned society. If so, score 1.
- It is conceivable that the researchers might be recommended for a Nobel Prize. If so, score 2.

10. Collaborators and commitment:
- The leading proposers have many irons in the fire. If so, score 0.
- The proposed research involves a new collaboration, but each collaborator has other interests. If so, score 1.
- New collaboration may or may not be involved, but the proposed research is a major interest for everyone involved. If so, score 2.

If your research would have a low BRAVERI, it is very unlikely to deserve a Blue Skies appellation. However, I do not imply that research with a low BRAVERI score cannot be innovative or adventurous. In the past, what might have been called mainstream research was the source of all points of departure. All the major discoveries stemmed from it. But scientists working in the mainstream today are rarely free to follow the unexpected leads they may come across "to fortune and fame unknown." In any case, mainstream research is readily funded by the conventional agencies. All our Venture Research programs would have scored more than 50%, although the index was not used at the time. I derived it as an amusement while waiting in the outer office of a possible sponsor as a send-up on today's fashion for quantification. The index will not be used in the future. It is given here for guidance only, and to hint at some of the beauty and excitement that lies in Blue Skies Research worthy of the name.

Sadly, the higher the BRAVERI, the less likely it is that the research would be funded by a national agency. *Almost all research funded today would have a BRAVERI close to zero.* That is a remarkable fact in itself. Our Venture Research initiative is searching for funds to support research that would have a high BRAVERI. Agencies claiming to support Blue Skies Research initiatives, and the governments and other investors who sponsor them, might care to apply our tongue-in-cheek BRAVERI assessment. They might be surprised at the outcome.

In conclusion, therefore, our message is that fostering an environment that stimulates and supports Blue Skies Research *as well as doing all the other things* will ensure that governments and industry have a reliable supply of genuinely new investment options. We would all be much richer intellectually. The genuinely new technology thereby generated would lead to increased economic growth. Research sponsors should be eager to extend their support. The changes required to the present arrangements to bring that environment about are marginal in terms of funds, but attitudes need to change too, and that is perhaps the most difficult thing to achieve.

The Venture Research scheme has sometimes been criticized as elitist, and so it is, but only in respect of the exceptional quality of its ideas. It strives to ignore power, influence, privilege, and fashion. Outstanding ideas do not grow on trees, and at any one time, very few scientists would expect to be working on one. That would be true most of the time even for the best scientists at the most prestigious institutes anywhere. Nevertheless, any competent and determined researcher might expect to have an inspirational idea at least once during the course of a lifetime. The problems then would come thick and fast. First, you would have to have the courage to do something about it. Then you would have to find an agency that will give you the unfettered backing and encouragement you need. As things stand at present, you would probably search in vain. Since none of us know when that inspired moment will come, it is therefore in every scientist's interests that the necessary modest changes are made as soon as possible.

Venture Research is one proven way of making those changes. As our international group of supporters said in a letter to *Nature* and *Science* (see Appendix 2), the additional costs would be tiny—about 0.04% of current spending on exploratory research. The major problem, however, is to raise a fund that would be free of the endless streams of restrictions and controls—committees, peer review, priority areas, key issues, milestones, and so on—every scientist must now routinely deal with. This may sound revolutionary, but all we want to do is to restore the freedom that was the norm not so long ago. Otherwise, global growth will suffer.

Business cycles normally ebb and flow, but we fear that the progressively deeper recessions of successive business cycles will eventually become so severe that recovery may not be possible. Civilization itself could be threatened. The letter was rejected without explanation. However, a chance meeting with Anjana Ahuja, a journalist from *The Times* of London, led to her writing a piece about the unpublished letter, entitled "Where is the next Einstein?" It was published in *The Times* on February 26, 2001, and attracted so much interest that it was eventually syndicated throughout much of the English-speaking world. The resultant correspondence led, astonishingly, to a spectacularly generous offer from a Middle Eastern group to provide us with our fund. "We will make your dreams come true," the group's representative told me. Unfortunately, the discussions came to an abrupt halt following the tragic events of September 11, 2001.

Our international group of supporters is growing steadily. We still live in hope.

Appendix 1

Some Results from the Venture Research Initiative Sponsored by British Petroleum (BP), 1980–1990

Venture Research was distinctive and successful because of its unconventional selection procedures. As was explained in Chapter 8, we in effect allowed Venture Researchers to select themselves, as we shall do in the future when we raise a new fund. We needed unconventional procedures because we were deliberately looking for those scientists funding agencies usually reject outright. Their ideas, as yet untested of course, would simply be too revolutionary. Nevertheless, such people are not easy to find, and it took some five frustrating years before we really knew what we were

Pioneering Research: A Risk Worth Taking, By Donald W. Braben
ISBN 0-471-48852-6   © 2004 John Wiley & Sons, Inc.

doing. The process is based on extensive face-to-face dialogue with researchers rather than on the opinions of their peers, and an exclusive focus on scientific concepts rather than possible products or processes. It now works well.

Each of the 26 groups participating in the scheme in 1990 had failed to secure adequate funds for their radical research from the usual sources simply because their peers did not agree that their contribution was needed or would be worthwhile or successful. Nevertheless, most, if not all, went on to achieve substantial scientific success, and a smaller number also went on to make a great deal of money.

Groups are listed alphabetically in their present (2003) locations:

- **Mike Bennett and Pat Heslop-Harrison (Kew Gardens and Leicester)** were the first to realize the significance of a three-dimensional structure within the cell nucleus. Thus, a gene's behavior—whether or not it is expressed, for example—is influenced by that gene's three-dimensional environment. They also discovered some of the rules governing the regulation and behavior of suites of genes as genomes evolve and hybridize with time. These changes can be substantial and give heritable changes independent of mutations.

- **Paul Broda et al. (University of Manchester Institute of Science and Technology)** pioneered the coherent study of the bacterial and fungal genes and enzymes that recycle the largest durable component of plant material (lignocelluloses). This group found new ways of triggering the expression of genes involved in lignin attack and enzymes that degrade lignin in grasses and straw. Normally, livestock animals are unable to extract the nutrients which lignin encapsulates, and so these techniques could considerably improve the nutritional value of traditional animal feeds.

- **Terry Clark et al. (Sussex)** demonstrated that overtly quantum-mechanical behavior could be associated with macroscopic objects. Their discovery will probably transform the design and scope of electronic circuitry in the long term as all electronic systems are currently based on classical or semiclassical science. The group has also developed new electronic techniques for characterizing weak electronic signals with high spatial resolution. These techniques have a wide range of applications in the short term, particularly in the biological fields.

- **Adam Curtis and Chris Wilkinson (Glasgow)** pioneered the direct study of electrical signal exchange between nerve cell networks and microfabricated external electrodes. The work developed slowly because of unexpected difficulties finding the best electrode system, but now it is possible to carry out a two-way conversation with small groups (four to five) of organized nerve cells. This is starting to give new understanding of the grammar and syntax of nerve cell language.

- **Steve Davies (Oxford)** identified highly efficient ways ($\sim 100\%$) of synthesizing chirally pure organic molecules. The work led to the establishment of a new company—Oxford Asymmetry—that was quoted on the London Stock Exchange. Its initial value (1998) was some £200 million.

- **Edsger W. Dijkstra (Texas at Austin), Netty van Gasteren (Eindhoven), and Lincoln Wallen (Oxford)** worked to improve, by an order of magnitude, upon the state of the art of presentation and systematic design of algorithms and mathematical proofs. Such an improvement was needed to make program design more reliable. A clear presentation style emerged that couples completeness with conciseness and pays ample attention to the why and how of the design decisions leading to the ultimate proof or program. The work extended over 12 years and made a substantial impact on BP's computer-based capabilities. Sadly, Edsger and Netty are now deceased.

- **Peter Edwards and David Logan (Oxford)** sought to understand electronic phase transitions such as between metals and nonmetals and to explore transitions to a new insulating phase consisting of atoms possessing permanent electric dipole moments. They found that the dipolar state is more widespread than hitherto had been thought. The new state may play a substantial role in understanding the behavior of high-temperature superconductors.

- **Nigel Franks (Bristol) and Jean Louis Deneubourg (Université Libre de Bruxelles)** were the first to begin quantifying the rules governing distributed intelligence in ant colonies. Because of their work, ant colonies are now seen as ideal model systems for understanding information flow and decision making in biological organizations in general. Their rule-based algorithms have been applied to both computer networks and the control of teams of robots.

- **Dudley Herschbach (Harvard)** pioneered the use of dimensional scaling as a route to calculating the electronic structure of atoms and molecules. This approach involves treating the dimension of space as a variable, and Herschbach found that it led to remarkable simplifications in calculating systems that are arduous or intractable using conventional techniques. Recently, the new technique has also proved effective in treating boson correlations in the burgeoning field of Bose–Einstein condensates.

- **Jeff Kimble (California Institute of Technology)** studied the quantum dynamics of optical systems and, among other things, pioneered new ways of producing squeezed quantum states of light. His group has also been at the forefront of the development of the new field of cavity quantum electrodynamics, an area of research that allows nonlinear optics to be performed at the level of single atoms and photons.

- **Graham Parkhouse (Parkhouse Consultants)** was the only industrial scientist to be supported by Venture Research. (His research was done at the University of Surrey in a post we arranged for him.) As a "theoretical engineer," he developed a new integrated approach to the theory of structures and, among other things, showed that brittleness can be an advantageous property in structural design. He also devised new ways of describing and controlling performance that could have a major impact on the design of new materials and composites.

- **Alan Paton, Anne Glover, and Eunice Allan (Aberdeen)** pioneered the development of novel symbioses between bacteria and plants. According to convention, it is not possible to engineer artificial associations between eukaryotic and prokaryotic organisms. However, some plant compartments (organelles) such as the chloroplast and the mitochondria would appear to be of bacterial origin. It is possible, therefore, that the group may have discovered a form of genetic exchange that may have been important in the evolution of the life forms that we see today. Their discoveries open up a wide range of scientific and technological possibilities, including the use of plants as bioreactors and the development of disease resistance, without recourse to traditional genetic engineering techniques. The full implications of their discoveries have barely been examined. Sadly, Alan is now deceased.

- **Martyn Poliakoff (Nottingham)** pioneered the use supercritical fluids as an environment for reaction chemistry. His work has led to a greater understanding of these unusual fluids and to the development of completely new industrial processes—a 1000-tons-pa plant is currently (2001) under construction at Consett, Co., Durham.

The supercritical environment can be fine tuned by varying temperature and pressure to favor certain chemical reactions and different ranges of products. Thus, these novel fluids may offer important advantages in the use of biofeedstock—plant materials rather than derived fossil fuels—as a source of organic chemicals such as plastics.

- **Alan Rayner et al. (Bath) and Ian Ross et al. (Santa Barbara, California).** Initially, these two groups were totally independent, but gradually they came to realize that they had common objectives. They recognized the potential role of mitochondria in regulating cell behavior a decade before the current paradigm of mitochondrial control of programmed cell death (apoptosis) and other senescence programs was accepted. They have expanded their hypothesis to include a mitochondrial role in regulating specific nuclear gene expression, which if perturbed could lead to cell and organ senescence and possibly explain the apparent random nature of causes of death, even among close siblings and clones.

- **Peter Rich (University College London),** as a postdoctoral researcher at Cambridge, launched a multidisciplinary attack on photosynthesis. (Conventionally, postdoctoral researchers are not usually allowed to be principal investigators.) He found a physical chemical understanding of how biological quinones function and went on to elucidate the mechanisms of several key enzymes involved in biological energy provision.

- **Ken Seddon (Queen's University of Belfast)** discovered the vast potential of performing chemistry in an ionic environment, as compared to, say, an aqueous or organic one. The scope of ionic liquids is vast, and Seddon realized that one could design ionic environments to optimize particular chemical reactions, a degree of freedom which is not open to those who use covalent solvents. His work has turned out to have huge industrial significance for developing novel "green" processes. The possibilities include nonpolluting alternatives to conventional solvents such as sulfuric and hydrochloric acid as well as solvents banned by the Montreal protocol.

- **Colin Self et al. (Newcastle)** studied the role of instability in biological systems. The processes of recognition and response in antibody–antigen systems are distinct. Thus, recognition need not imply a response because the antigen may decay before the cell can activate its commitment. This group has extended our understanding of biological control systems considerably. Among other things, they were the first to photoactivate an antibody, a vital step toward antibody-directed therapy in medicine.

· **Gene Stanley (Boston) and José Teixeira Laboratoire Leon Brillouin (CEA-CNRS, France)** studied the properties of water in confined geometries. When water occupies small volumes, as in porous materials, or when it binds to biological materials, the hydrogen bond network is modified. Consequently, the structural and dynamic properties of confined water are different from those in the bulk state. The group discovered a second critical point in liquid water, which feature explains the long puzzling anomalies found experimentally.

· **Harry Swinney (Texas at Austin) and Patrick DeKepper et al. (Center de Recherche Paul Pascal, Bordeaux).** Conventionally, chemical reactions occur in well-stirred reactors in an environment that is essentially zero dimensional. This group studied self-organization in chemical reactions and progressed into the regime of multidimensional chemistry. Thus, they were the first to produce the chemical spatial patterns predicted in a classical paper by Alan Turing in 1952. They were also the first to study the space–time evolution of patterns—the spatial distribution of reaction products—in chemical systems.

· **Robin Tucker et al. (Lancaster)** pioneered a mathematical reformulation of inherently nonlinear phenomena offering new avenues within gravitation and quantum field theory. The unexpected outcome was that they also made important contributions to other fields, including spinning strings, membrane theory, and the unification of the fundamental interactions. In addition, their work, which might normally be regarded as of archetypal academic purity, has been found to be relevant to a wide range of industrial problems, for example, the stability of bridges under wind and rain excitations, friction forces on rotating drill strings, and fatigue damage to undersea marine risers due to the periodic emission of vortices.

Appendix 2

The Venture Research Group

An international group of senior scientists has agreed to join a Venture
Research Group. On January 15, 2001, the group sent the following letter
to the editor of *Nature*. He declined to publish it. A few days later, it was
sent to the Letters Editor of *Science*, who also declined. No reasons were
given.

Dear Sir,
Max Planck made his seminal discovery of the quantisation of energy a
century ago. Within 25 years, our understanding of matter and radiation
had been revolutionized. After another few decades, the DNA code had
been revealed, superconductivity explained, and the transistor and laser
invented. These incredible feats transformed civilization. They were the
products of industrial and academic research environments that once
thrived on individual freedom. These environments are now under severe
threat.

All too often today, the academic research environment favors objectives
selected by consensus. Adequate provision must be made for work in the

Pioneering Research: A Risk Worth Taking, By Donald W. Braben
ISBN 0-471-48852-6 © 2004 John Wiley & Sons, Inc.

mainstreams, of course, but current policies make it almost impossible for latter-day Plancks to flourish. We believe that this is one of the most serious problems facing civilization.

The problem can be solved by an adjustment to the global funding of basic research—currently about $50 billion pa. Some $10–20 million pa (or 0.04% of the total) should be sufficient to germinate the rare seeds of the Planck variety. But how can they be reliably identified? Pioneers and consensus can be poor bedfellows initially, and so peer-review often fails. There is an urgent need, therefore, for efficient selection procedures that are more tolerant of risk. We the undersigned intend therefore to create a Forum of concerned scientists, engineers, and others to draw attention to the problem. It would explore ways that a Planck-type fund might be raised, and stimulate the exploration of uncharted territories.

Don Braben*
Venture Research International Ltd
Mount End
Theydon Mount
Epping
Essex CM16 7PS,
and, University College London, UK.
DonBraben@compuserve.com
And:
Mike Bennett, Jodrell Laboratory, Kew Gardens
Terry Clark, University of Sussex
Peter Cotgreave, Save British Science Society, London
John Dainton, University of Liverpool
Peter Edwards, University of Birmingham
Nigel Franks, University of Bristol
John Guest, University College London
Dudley Herschbach, Harvard University
Jeff Kimble, California Institute of Technology
Harry Kroto, University of Sussex
David Price, University College London
David Ray, Oxford Innovation Ltd
Ian Ross, University of California at Santa Barbara
Ken Seddon, The Queen's University of Belfast
Gene Stanley, Boston University
Harry Swinney, University of Texas at Austin
Robin Tucker, University of Lancaster
Luca Turin, University College London
Claudio Vita-Finzi, University College London
23 January 2001

*To whom correspondence should be addressed.

Bibliography

Abbate, E., et al. (1998). *Nature*, June 4, p. 458.

Abramovitz, M. (1991). *Thinking about Growth*. Cambridge University Press, Cambridge.

Arrow, K. (1962). Economic Welfare and the Allocation of Resources for Invention. In *The Rate and Direction of Inventive Activity*. Princeton University Press, Princeton, New Jersey.

Barnet, C. (1986). *The Audit of War*. MacMillan.

Barro, R. J., and Xavier Sala-i-Martin. (1995). *Economic Growth*. McGraw-Hill, New York.

Braben, D. W. (1994). *To Be a Scientist*. Oxford University Press.

Brenner, S. (1998). *Science*, November 20, p. 1411.

Brewer, J. (1989). *The Sinews of Power*. Unwin Hyman, London.

Butterfield, H. (1958). *The Origins of Modern Science 1300–1800*. Bell and Sons, London.

Cardwell, D. S. L. (1972). *Technology, Science and History*. Heinemann, London.

Chargaff, E. (1978). *Heraclitean Fire*. Rockefeller University Press, New York.

Clark, R. W. (1961). *The Birth of the Bomb*. Phoenix House, London.

Clark, R. W. (1962). *The Rise of the Boffins*. Phoenix House, London.

Collier, P., and Hoeffler, A. (2003). *Greed and Grievance in Civil War*. World Bank.

Davies, N. (1996). *Europe*. Oxford University Press, Oxford.

de Solla Price, D. (1975). *Science Since Babylon*. Yale University Press, New Haven, Connecticut.

Pioneering Research: A Risk Worth Taking, By Donald W. Braben
ISBN 0-471-48852-6   © 2004 John Wiley & Sons, Inc.

Diamond, J. (1997). *Guns, Germs and Steel.* Jonathan Cape, London.

Donahue, W. H. (1992). *The New Astronomy* (translation of Kepler's magnum opus). Cambridge University Press, Cambridge.

Doolittle, W. F. (2001). *Science*, March 2, p. 1707.

Drucker, P. (2001). The Next Society: A Survey of the Near Future. *The Economist*, November 3.

Ehrlich, P. (2000). *Human Natures: Genes, Cultures, and the Human Prospect.* Island.

Eltis, W., and Sinclair, P. (Eds.). (1988). *Keynes and Economic Policy.* MacMillan.

Engineering and Physical Sciences Research Council (EPSRC). (1998, May). *Action for Foresight.* EPSRC.

Faraday, M. (1900?). *Experimental Researches in Electricity.* J. M. Dent & Sons, London.

Feiling, K. (1966). *History of England.* Oxford University Press.

Fernandez-Armesto, F. (1995). *Millennium.* Bantam Press, London.

Feynman, R. (1989). *What Do You Care What Other People Think?* Unwin Hyman, London.

Finley, M. I. (1963). *The Ancient Greeks.* Penguin Books.

Galbraith, J. K. (1952). *American Capitalism.*

Gibbon, E. (1796). *The Decline and Fall of the Roman Empire.*

Goldman, M. A. (2001). *Nature*, September 20, p. 252.

Hall, A. R. (1970). *From Newton to Galileo.* Fontana-Collins.

Hartley, Sir H. (1972). *Humphry Davy.* EP Publishing, Wakefield.

Herodotus. (1996). *The Histories* (translated by A. de Sélincourt, 1972, revised by J. Marincola). Penguin, London.

Hill, C. (1965). *Intellectual Origins of the English Revolution.* Clarendon, Oxford.

Jewkes, J., Sawyers, D., and Stillerman, R. (1969). *The Sources of Invention.* MacMillan.

Kennedy, P. (1993). *Preparing for the Twenty-First Century.* HarperCollins, London.

Klemm, F. (1959). *A History of Western Technology.* George Allen and Unwin, London.

Koestler, A. (1964). *The Sleepwalkers.* Penguin Books.

Lederman, L. (1991). *Science: The End of the Frontier?* American Association for the Advancement of Science.

Lewin, R. (1996). *Lise Meitner. A Life in Physics.* University of California Press.

Lodge, O. (1893). *Pioneers of Science.* MacMillan, London.

Maddison, A. (1989). *The World Economy in the Twentieth Century.* Organisation for Economic Co-operation and Development, Paris.

Maddison, A. (1995). *Monitoring the World Economy 1820–1992.* Organisation for Economic Co-operation and Development, Paris.

Masson, I. (1935). *Science Today.* Eyre & Spottiswood, London.

Mellars, P. (1998). *Nature,* October 8, p. 539.

Mill, J. S. (1848). *Principles of Political Economy.*

Monod, J. (1972). *Chance and Necessity.* Collins, London.

Morgan, L. H. (1877, reprinted 1964). *Ancient Society.* Harvard University Press, Cambridge, Massachusetts.

Mowery, D. C., and Rosenberg, N. (1989). *Technology and the Pursuit of Economic Growth.* Cambridge University Press, Cambridge.

National Science Foundation (NSF). (1999). Survey of Doctorate Recipients. NSF, Washington, D.C.

Nelson, R. (1959). *The Simple Economics of Basic Research.* Published in *The Economics of Technological Change,* N. Rosenberg (Ed.), Penguin, 1971.

Nelson, R. R., Peck, M. J., and Kalachek, E. D. (1967). *Technology Economic Growth and Public Policy.* Brookings Institution, Washington, D.C.

Peck, L. L. (1990). *Court Patronage and Corruption in Early Stuart England.* Unwin Hyman, London.

Plumb, J. M. (1950). *England in the Eighteenth Century.* Penguin Books.

Polanyi, J. (1995). *Daily Telegraph,* April 2.

Press, F. (1988). "The Dilemma of the Golden Age," Address to members at 125th Annual Meeting of the National Academy of Sciences, April 26.

Raby, P. (2001). *Alfred Russel Wallace: A Life.* Chatto and Windus.

Rees, M. (2003). *The Final Century.* William Heineman, London.

Roberts, C. (1966). *The Growth of Responsible Government in Stuart England.* Cambridge University Press.

Rosenberg, N. (Ed.). (1971). *The Economics of Technical Change.* Penguin Books.

Rowse, A. L. (1950). *The England of Elizabeth.* Macmillan.

Smith, A. (1776). *An Inquiry into the Nature and Causes of the Wealth of Nations.*

Smith, D. K., and Alexander, R. C. (1988). *Fumbling the Future.* William Morrow and Company, New York.

Soros, G. (1998). *The Crisis of Global Capitalism.* Little Brown, London.

Stanley, H. E., et al. (1999). *Nature,* July 29, p. 437.

Starr, C. G. (1983). *A History of the Ancient World.* Oxford University Press, New York.

Stillman, J. M. (1924). *The Story of Early Chemistry.* D. Appleton and Company, New York.

Turner, R. (1941). *The Great Cultural Traditions.* McGraw-Hill, New York.

Waddington, C. H. (1977). *Tools for Thought.* Paladin, 1977.

Wang, J. (1999). *American Scientists in an Age of Anxiety.* University of North Carolina Press.

Ziman, J. (1994). *Prometheus Bound.* Cambridge University Press.

Index

Note: references to figures and tables are indicated as 5f, 100f; to footnotes are indicated as 5n, 100n; references to illustrations and captions as 5i, 100i; to the Text Boxes as 5T, 100T.

Pioneering Research: A Risk Worth Taking, By Donald W. Braben
ISBN 0-471-48852-6   © 2004 John Wiley & Sons, Inc.

diseases, 12, 39, 137
  resistance, 139T
dissent, 1–5, 7, 83, 144
  in evolution, 10–11, 13, 14, 16–21, 27–28,
    31
  in history, 36–37, 44, 53, 56
DNA, 69, 90
Doolittle, W. Ford, 91T
Dowland, John L., 91T
Drucker, Peter, 121
DuPont, 129

East India Company, 120–121
economic growth *see* growth, economic
Economist, The, 39n, 94, 98, 112–113, 115,
    119, 120, 121, 126, 141
Edison, Thomas, 128–129
education, 45, 47, 57, 124
  ancient, 35–36, 37, 38–39
  see also universities
Edwards, Peter, 181
efficiency, 3, 5, 7, 8, 13, 78, 83, 85, 92, 107–
    108, 114, 120, 121
Egypt, ancient, 35
Ehrlich, Paul, 31n
Einstein, Albert, 83, 90, 132T
elements, the four, 47
Elizabeth I, Queen, 52n, 53
empiricism, 36n
Engineering and Physical Sciences Research
    Council (EPSRC), 71, 74, 79, 112n
Entertaining Science lectures, 167n
environment, 6, 142, 144
  pollution, 142, 143T
environmentalists, 140, 142, 144
equilibrium, state of, 143–144
equipment, 102, 104, 135, 159, 163
Euclid, 36
European Science Foundation, 89
European Union (EU), 119
evolution, 9–10, 14–20, 26
  and the Church, 27T
*Excellence and Opportunity: A Science
    Policy for the 21st Century,* 77–78

Fairclough, Sir John xi, 172
Faraday, Michael, 82
Feiling, Keith, 42n
Feshbach, Murray, 39n
Feynman, Richard P., 66–67

Financial Times, 171
Finlay, M. I., 36n
Foresight *see* Technology Foresight
fractal geometry, 85, 106T
Franklin, Benjamin, 58
Franks, Nigel, 181
freedom, 12, 110–111, 122, 138, 152T, 156–
    157
Frisch, Otto, 132T
fund raising, 173, 177–178
funding, research, 5, 60–63, 67, 69, 70, 71,
    90–92, 101–104, 123, 134, 140, 145
  agencies, 85, 103–104, 160, 162
  ancient Greece, 35
  pre-*1970*s, 104–105, 108
  free, 162–163
  global, 186
  HEFCE, 79
  patrons, 34, 49, 60, 66T
  universities, 79–86, 90, 101–102, 104–105
  US, 91–92, 101–102
  VRU, 90, 158–159, 162–163
  see also grants, research
funding councils, 79–81

Galileo, 47, 50, 51T
game theory, 89T
GDP *see* gross domestic product
Gell-Mann, Murray, 19
General Electric, 110T, 128n
genetics, 133–135, 180
  genetically modified foods, 141–142
  jumping genes, 139T
genomes, 133–135, 142, 180
Gibbon, Edward
  *The Decline and Fall of the Roman
    Empire,* 38T, 40
Glaucos of Chios, 32T
Glover, Anne, 182
Golden Age (*1950–73*), 12, 58n, 59–60,
    109–110, 127, 129
Goldman, Michael A., 18n
Grant Proposal Guide (NSF), 91
grant-awarding bodies, 66T
grants, research, 71–72, 80–81, 91–92
  NSF, 91
  university, 80T
  see also funding; research proposals
gravity, 22
Greek language, 38, 42